Computational Intelligence Techniques and Their Applications to Software Engineering Problems

Computational Intelligence Techniques and Their Applications to Software Engineering Problems

Edited by
Ankita Bansal, Abha Jain, Sarika Jain,
Vishal Jain and Ankur Choudhary

CRC Press is an imprint of the
Taylor & Francis Group, an **informa** business

First edition published 2021
by CRC Press
6000 Broken Sound Parkway NW, Suite 300, Boca Raton, FL 33487-2742

and by CRC Press
2 Park Square, Milton Park, Abingdon, Oxon, OX14 4RN

CRC Press is an imprint of Taylor & Francis Group, LLC

ISBN: 978-0-367-52974-1 (hbk)
ISBN: 978-1-003-07999-6 (ebk)

Typeset in Times
by codeMandra

Contents

Preface

No one can imagine our modern life without software. There is a huge demand for quality software with fast response and high reliability. This increasing demand is bringing various challenges for software industries. Before deployment, software undergoes various developmental stages such as eliciting requirements, designing, project planning, coding, testing and maintenance. Every stage is bundled with numerous tasks or activities. The ever-changing customer requirements and the complex nature of software make these tasks more challenging, costly and error-prone. To overcome these challenges, we need to explore computational intelligence techniques for different software engineering tasks.

Computational intelligence is closely related to artificial intelligence in which heuristic and metaheuristic algorithms are designed to provide better and optimized solutions at reasonable costs. Computational techniques, such as optimization techniques, metaheuristic algorithms and machine learning approaches, constitute different types of intelligent behavior. Optimization techniques are approaches that provide optimal or near-optimal solutions to problems where goals or targets to be achieved are known. Metaheuristic is a high-level, iterative process that guides and manipulates an underlying heuristic to efficiently explore the search space. The underlying heuristic can be a local search or a low- or high-level procedure. Metaheuristics provide near-optimal solutions with high accuracy and limited resources in a reasonable amount of time by exploiting the search space. Machine learning algorithms work efficiently when we have sufficient data to extract knowledge and train models. For example, models can be developed to classify error-prone classes of software. These algorithms have proven their effectiveness in different application domains such as medicine, bioinformatics, computer networks and weather forecasting. Researchers have applied computational intelligence techniques to solve various problems in the software engineering domain such as requirement prioritization, cost estimation, reliability assessment, defect prediction, maintainability prediction, quality prediction, size estimation, vulnerability prediction and test case prioritization, among others.

In software industries, the stakeholders of software organizations conduct daily activities such as software designing, project planning and testing, among others. Computational intelligence techniques can be applied to carry out these activities efficiently. In addition to application in software industries, computational intelligence techniques can have real-life applications, such as in household appliances and medicine (e.g., tumor identification, disease diagnosis, X-ray, etc.). The aim of *Computational Intelligence Techniques and Their Applications to Software Engineering Problems* is to focus on the application of computational intelligence techniques in the domain of software engineering. In this book, researchers and academicians have contributed theoretical research articles and practical applications in the field of software engineering and intelligent techniques.

This book would be primarily useful for researchers working on computational intelligence techniques in the field of software engineering. Moreover, because this book provides a deep insight into the topic from diverse sources, beginner and intermediary researchers working in this field would find this book highly beneficial.

We express our sincere thanks to Gagandeep Singh, publisher (engineering), CRC Press, for giving us an opportunity to convene this book in his esteemed publishing house, and Lakshay Gaba, editorial assistant, CRC Press, for his kind cooperation in completion of this book. We thank our esteemed authors for having shown confidence in this book and considering it as a platform to showcase and share their original research. We also wish to thank the authors whose research was not published in this book probably because of minor shortcomings.

Editors

Ankita Bansal is an assistant professor at Netaji Subhas University of Technology (NSUT), Delhi, India. Prior to joining NSUT, Dr. Bansal worked as a full-time research scholar at Delhi Technological University (DTU) (formerly Delhi College of Engineering). She received her master's and doctoral degrees in computer science from DTU. Her research interests include software quality, soft computing, database management, machine learning and metaheuristic models.

Abha Jain is an assistant professor at Shaheed Rajguru College of Applied Sciences for Women, Delhi University, India. Prior to joining the college, she worked as a full-time research scholar and received a doctoral research fellowship from DTU. She received her master's and doctoral degrees in software engineering from DTU. Her research interests include data mining, software quality and statistical and machine learning models. She has published papers in international journals and conferences.

Sarika Jain graduated from Jawaharlal Nehru University (India) in 2001. She has served in the field of education for over 19 years and is currently working at the National Institute of Technology Kurukshetra, India. Dr. Jain has authored/coauthored over 100 publications including books. Her current research interests include knowledge management and analytics, semantic web, ontological engineering and intelligent systems. Dr. Jain has supervised two doctoral scholars (five ongoing) who are now pursuing their postdoctorates. She has two research-funded projects: one ongoing project is funded by CRIS TEQUIP-III, and the other completed project is funded by DRDO, India. She has also applied for a patent. Dr. Jain has been supervising DAAD interns from different German universities and works in collaboration with various researchers across the globe including Germany, Austria, Australia, Malaysia, the United States, Romania and many others. She is a member of IEEE and ACM and is a Life Member of CSI, IAENG and IACSIT.

Vishal Jain is an associate professor at the Bharati Vidyapeeth's Institute of Computer Applications and Management (BVICAM), New Delhi, India (affiliated with Guru Gobind Singh Indraprastha University and accredited by the All India Council for Technical Education). He first joined BVICAM as an assistant professor. Prior to that, for several years he worked at the Guru Premsukh Memorial College of Engineering, Delhi, India. He has more than 350 research citations and has authored more than 70 research papers in reputed conferences and journals including *Web of Science* and *Scopus*. Dr. Jain has authored and edited more than 10 books with various reputed publishers including Springer, Apple Academic Press, Scrivener, Emerald and IGI-Global. His research areas include information retrieval, semantic web, ontology engineering, data mining, adhoc networks and sensor networks. He was a recipient of the Young Active Member Award for 2012–2013 from the Computer Society of India, Best Faculty Award for 2017 and Best Researcher Award for 2019 from BVICAM, New Delhi.

Ankur Choudhary majored in computer science and engineering and pursued a PhD from the School of ICT, Gautam Buddha University (GBU), Greater Noida, India. He has more than 15 years of teaching experience and is currently a professor at Sharda University, Greater Noida, India. His areas of research include nature-inspired optimization, artificial intelligence, software engineering, medical image processing and digital watermarking. Dr. Choudhary has published research papers in various SCI/Scopus-indexed international conferences and journals and is associated with various international journals as a reviewer and editorial board member. He has organized conferences and special sessions at international conferences and has served and delivered as session chair.

Contributors

Ashish Agrawal
Department of Computer Science
SRMS College of Engineering & Technology
Bareilly, India

Monika Arora
Department of Computer Science and Engineering
Apeejay School of Management
Dwarka, India

Anu Bajaj
Department of Computer Science and Engineering
Guru Jambheshwar University of Science and Technology
Hisar, India

Shivani Bali
Department of Operations Management & Business Analytics
Lal Bahadur Shastri Institute of Management
Delhi, India

Vikram Bali
Department of Computer Science and Engineering
JSS Academy of Technical Education
Noida, India

Mamta Bansal
Department of Computer Science and Engineering
School of Engineering & Technology, Shobhit Deemed University
Meerut, India

Pradeep Kumar Bhatia
Department of Computer Science and Engineering
Guru Jambheshwar University of Science & Technology
Hisar, India

Indu Chhabra
Department of Computer Science & Applications
Panjab University
Chandigarh, India

Jitender Kumar Chhabra
Department of Computer Engineering
National Institute of Technology
Kurukshetra, India

Mohamed Elwakil
Math, Physics & Computer Science Department
University of Cincinnati Blue Ash College
Blue Ash, Ohio

Murtaza Mohiuddin Junaid Farooque
Department of Management Information Systems
Dhofar University
Salalah, Oman

Sakthivel Gnanasekaran
Multidisciplinary Centre for Automation, School of Mechanical Engineering
Vellore Institute of Technology Chennai
Chennai, India

Somya Goyal
Department of Computer and Communication Engineering
Manipal University Jaipur
Jaipur, India
Department of Computer Science and Engineering
Guru Jambheshwar University of Science & Technology
Hisar, Haryana

Deepti Gupta
Department of Information Technology
Institute of Innovation in Technology and Management
New Delhi, India

Anju Khandelwal
Department of Mathematics
SRMS College of Engineering & Technology
Bareilly, India

Maushumi Lahon
Department of Computer Engineering
Assam Engineering Institute School of Technology
Guwahati, India

Sushma Malik
Department of Computer Science
Shobhit Institute of Engineering and Technology (deemed-to-be-university)
Meerut, India

Anshu Ohlan
Department of Mathematics
Pt. Neki Ram Sharma Government College
Rohtak, India

Naveen Kumar P.
School of Mechanical Engineering
Vellore Institute of Technology Chennai
Chennai, India

Minakshi Rajput
National Institute of Electronics and Information Technology
Dr. Babasaheb Ambedkar Marathwada University
Aurangabad, India

Narayana Swamy Ramaiah
Department of Computer Science and Engineering
Jain University (deemed-to-be-university)
Bangalore, India

Anamika Rana
Department of Computer Science and Engineering
Maharaja Surajmal Institute of Technology
New Delhi, India

Amit Rathee
Department of Computer Engineering
National Institute of Technology
Kurukshetra, India

Wasiur Rhmann
Department of Computer Science and Information Technology
Babasaheb Bhimrao Ambedkar University (A Central University)
Amethi, India

Santhosh S.
Department of Computer Science and Engineering
Jain University (deemed-to-be-university)
Bangalore, India

Ganesh Sable
Department of Electronics and Telecommunication
Maharashtra Institute of Technology
Aurangabad, India

Neha Saini
Department of Computer Science & Applications
Panjab University
Chandigarh, India

Om Prakash Sangwan
Department of Computer Science and Engineering
Guru Jambheshwar University of Science and Technology
Hisar, India

Uzzal Sharma
Department of Computer Science and Engineering & Information Technology
Assam Don Bosco University
Guwahati, India

Jitendra Singh
Department of Software Engineering
Wachemo University
Hosanna, Ethiopia

Gaurav Singhania
Department of Computer Science and Engineering
JSS Academy of Technical Education
Noida, India

1 Implementation of Artificial Intelligence Techniques for Improving Software Engineering

Sushma Malik
Shobhit Institute of Engineering and Technology

Monika Arora
Apeejay School of Management

Anamika Rana
Maharaja Surajmal Institute of Technology

Mamta Bansal
Shobhit Deemed University

CONTENTS

1.1 INTRODUCTION

Computer science is the main branch of science that has gone through speedy and essential transformation. This transformation is a result of mixing the available experience and user expectations for designing new creative software products and improving the lives of users by modifying the environment (Cook, Augusto, & Jakkula, 2009).

Software engineering (SE) is a branch of computer science. This branch of computer science follows methodical and scientific approaches for software design, develops software on the basis of requirements, implement and maintain the software, and eventually retire the software product after some period. Although this is a systematic and linear approach to software development, it faces some of the following problems:

- It's not possible to read the brain of human beings or their behavior by using SE.
- Computer awareness is impossible with SE.
- Nondeterministic Polynomial (NP)'s problems are not easy to solve with SE.
- SE product development models use sequential phases, which makes products static in nature, but software products are not dynamic in nature.
- Real-time software development is not possible for engineers to design and develop with SE.

Therefore, software development is still more a craft or art than an engineering discipline, because it lacks a validation process and product modification in the SE improvement process (Jain, 2011).

Artificial intelligence (AI) is the branch of computer science whose main motives are to develop intelligent machines, and define and design intelligent agents (Jain, 2011). AI focuses on creating machines that behave and think like human beings, creating a computer system that has some form of intelligence, and trying to implement that knowledge in understanding natural language (Aulakh, 2015).

AI and SE developed separately without much exchange of research results. The main goal of SE is to design efficient and reliable software products, while that of AI is to make the software product more intelligent. Today, these fields are coming closer and creating new research areas (Sorte, Joshi, & Jagtap, 2015), and many researchers are taking an interest in these fields. Many people are using Siri, Google Assistant, or Echo in their daily routine. What are these tools? In simple words, these devices are the personal digital assistants that help users access valuable information and basically work on the user's voice command like "Hey Siri, show me the closest petrol pump", "play some music" or even "call Ms. ABC." These digital supporters will provide appropriate information by going through the user's mobile (call or read the message) or searching the web. These are some examples of AI. The father of AI is John McCarthy, and according to him, "AI is the science and engineering of making intelligent machines, especially intelligent computer programs." AI is implemented using knowledge of how the human brain works and how human beings become skilled and make decisions to accomplish the steps in a task while trying to resolve a problem.

Hybridization of SE and AI is the most effective area of computer science and makes life easier for developers, testers, and analyzers. It is basically an opportunity to implement AI tools and approaches in the SE field to improve the software product development process. Software engineers have knowledge of AI technology and are interested in adopting the tools and techniques of AI in their own software development for improving the quality of the software product. This combination of fields is easily adopted and reused by nonexperts. AI will change the way an application or system is developed, and thus developers can expect a better application or system developed under the existing environment.

AI impacts all aspects of SE including data collection, software development, testing, deployment and ongoing maintenance. AI helps the developer to create better functioning software and increase the efficiency and reliability of software creation by automating manual tasks. AI enables machines to make decisions so that machines can balance compute power and service loads.

1.1.1 Literature Review

AI plays a significant role in automating numerous software development activities, and implementation of AI in SE makes sense to take advantage of AI techniques in the various phases of the Software Development Life Cycle (SDLC), like requirement gathering, designing, development, implementation, and testing (Kumari & Kulkarni, 2018; Sorte, Joshi, & Jagtap, 2015). Basu et al. (2017)analyzed views about the integration of AI and SE and explored the possibilities of combining the techniques and tools of AI in the designing and development of software systems. Using an AI automated programming tool in the development of software products eliminates the risk assessment phase and reduces the development time of software products. AI also increases the quality of the software product (Saini, 2016). SE helps design a software product, but development of the product is extremely time-consuming. Implementing AI techniques and tools in the development of software products increases the quality of software products and minimizes development time. The software development coding phase can be amended with the genetic algorithms (Aulakh, 2015). AI, expert systems, and knowledge engineering are playing an essential role in the development of software products. The combination of AI and SE is important; the advantages of AI techniques and approaches are used in SE and vice versa. The study by Shankari & Thirumalaiselvi (2014) suggests that the application of AI techniques can help in solving the problems that are associated with the SE process. Ammar, Abdelmoez, & Hamdi (2012) discussed how AI techniques are used to solve problems faced by software engineers in the software development phases, like requirement-gathering, design and coding and testing. They also summarized the problems like transforming the requirements into architectures using AI techniques. Harman (2012) explained that software is dynamic in nature and that its development process and deployment techniques always need modification based on requirements. AI is well suited to this task. Jain (2011) explained the interaction between SE and AI and the reason why AI techniques are needed in the SE process. Many AI techniques, like knowledge-based systems, neural networks, fuzzy logic (FL), and data mining, are used by software engineers and researchers to get better software products. AI plays an important role in improving all the phases of the SDLC (Meziane & Vadera, 2010). Risk can be reduced in development time by using AI automated tools. All the phases of software development are linked together and software product development time can be minimized by revisiting each and every phase after requirements are modified (Raza, 2009). Chen et al. (2008) explained the various AI techniques like case-based reasoning (CBR), rule-based systems, artificial neural networks (ANNs), fuzzy models, genetic algorithms (GAs), cellular automata, multiagent systems, swarm intelligence, reinforcement learning and hybrid systems with examples.

1.2 ASPECTS OF SE AND AI

The development of software products using well-defined sequential models and procedures is associated with the SE branch of computer science. SE is basically the combination of two words: *software* and *engineering. Software* is not just a programming code that is written in a programming language, but is a collection of executable codes written in some specific programming language which are used to develop a specific software product. However, *Engineering* is developing a product using well-defined scientific principles and methods. The process of developing a software product on the basis of SE principles and methods is called *software evaluation*, and it basically includes development, design, maintenance, and updates of the software product on the basis of the requirements. SE paradigms consist of all the engineering concepts that are used in software product development. The SDLC is the sequence of stages that are required to develop the software product. The main phases of SDLC include Requirement Gathering, Feasibility Study, System Analysis, Software Design, Coding Testing, Integration, Implementation, Operation and Maintenance, and finally, Disposition of the Software Product. Software can be developed and designed in any programming language. Then why is SE required? SE is necessary because user requirements are dynamic and the environment in which the software operates changes. SE can be classified into two categories:

1. Programming knowledge
2. Domain knowledge

Programming knowledge includes familiarity with the data structure construction, control structure, the syntax of programming languages, and much more to develop the software. However, domain knowledge basically includes the concepts, theories, and equations that characterize the particular field. A field category contains distinctive characteristics of the software programming language that is used to develop the software (Aulakh, 2015; Saini, 2016).

The traditional software development process starts with the collection of requirements from the user, and testing is the closing stage of the product. At each level, different kinds of knowledge are required; for example, in the design stage, designing knowledge is needed, and in the coding phase, programming and field knowledge are required (Figure 1.1).

The basic problem phase in SE is a long period of time between the collection and specification of requirements and the release of the software product. Sometimes requirements change before the release of the product, which causes the problem of phase independence; any changes in one phase of the software development process also affect subsequent levels. Therefore, the development team needs to modify the coding of the program when requirements are altered (Saini, 2016).

AI is implemented based on perceptions of how the human brain thinks and how a human being becomes skilled and makes a decision to accomplish a task in the steps necessary to resolve a problem. The outcome of this study is used in designing and developing software products and systems. AI is concerned with the study and creation of software systems with a particular aptitude and implements that knowledge in designing a system that can understand natural language. AI understands human perceptions, learning, and reasoning to solve complex problems. AI focuses on creating a system that can behave in a manner that human beings consider intelligent. AI tries to imitate human reasoning in a computer system. AI works with a large amount of data, and then smart algorithms are implemented in the software system to gain knowledge from the available data. AI is a vast subject in the computer science field and includes many advanced and complex processes. The study of this field has combined many theories, methods, and technologies. The major subfields of AI are machine learning, neural networks, deep learning, cognitive computing, and natural language processing.

The term AI was coined for the development of intelligent machines. The main goal of AI systems is to create a smart thing or product that can perceive its surroundings and react on the basis of those perceptions. AI is used to design expert systems that uses the knowledge of data instantly to control the process of solution. The expert system can be designed on the basis of knowledge gained from the expert, coding that knowledge in an appropriate form, validating the knowledge and ultimately designing a system with the help of AI tools (Raza, 2009) (Figure 1.2).

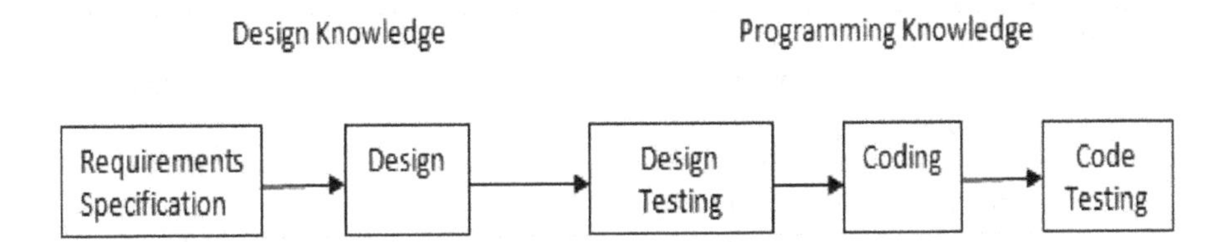

FIGURE 1.1 The traditional process of software development.

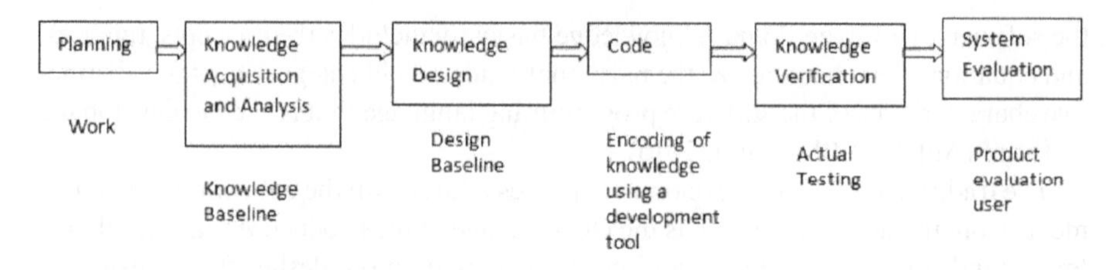

FIGURE 1.2 Expert system development.

1.2.1 Factors of Interaction between AI and SE

With the increased use of AI methods and techniques in the area of SE, the following factors are involved in the interaction between AI and SE:

a. **Objectives**: SE and AI have their own objectives and goals and engage in the development of software, but the goals of these fields are different. The main goals of SE are designing and developing a reliable, good-quality software by using specific approaches within the estimated time and cost constraints, whereas the goal of AI is development of intelligent software or machines (Jain, 2011).
b. **Problems**: SE and AI encounter different problems. How a problem is resolved is based on the factor that is involved and planning for the interaction between AI and SE. AI deals with problems for which solution methods are not well defined. To solve that problem, researchers in AI have designed and developed a number of powerful programming tools and techniques. However, SE deals with the problem of how to understand human behavior.
c. **Communication**: To implement both SE and AI in the development of software, you need to understand at what level the communication can be done between the two areas of specialization. The objective of SE is to develop and improve the quality of software, so the software can be developed within time and cost constraints, which involves issues like productivity, maintenance, reliability, and reuse of software products. In the same way, AI also has some objectives and issues, like how to solve the problem. So the level of communication between AI and SE is at the level of objectives and issues, not at the level of solution of the problem (Jain, 2011; Tsai, Heisler, Volovik, & Zua, 1988).
d. **Reasons for Implementing AI in the SE Field**: The main reasons for the implementation of AI methods, tools, and techniques in the SE field are as follows (Jain, 2011):
   - Automation programming in AI is similar to SE, and this represents the new model for SE research in the future.
   - AI expert systems are flourishing and offer significant solutions for some parts of software product development.
   - The development and maintenance environment of AI is suitable for the SE field.

- AI techniques and tools are used in SE for development of software products.
- Rapid development of prototype AI models is very effective in SE.

c. **Reasons for Not Implementing SE in the AI Field**: The main reasons that SE methods and tools are not used or implemented in the AI field are as follows (Jain, 2011):
   - SE uses a sequential approach for the development of software and has fixed phases, which are not suitable for AI because of its dynamic nature.
   - Understanding human behavior using SE is not easy.
   - The functioning of expert systems cannot specify the problems/requirements properly, so SE techniques cannot be applied in that field.
   - Maintenance of AI software is easy, so there is no need to use SE in the AI field.
   - The subfields of AI do not interact with each other, so it is not easy for them to communicate with SE (Figure 1.3).

1.2.2 Research Areas of Interaction between AI and SE

The AI and SE fields of computer science have developed separately in their research areas, and they do not share research results. Techniques of AI study make possible the software or product recognize, reasonable and intelligent. In another way, SE is helping software engineers to design and develop better software in less time. The relationship between AI and SE is minimal, but both fields are developing and growing individually. The techniques and methods of one field have emerged to another field. Both fields deal with real-world objects like business processes, expert knowledge, and process models (Sorte, Joshi, & Jagtap, 2015).

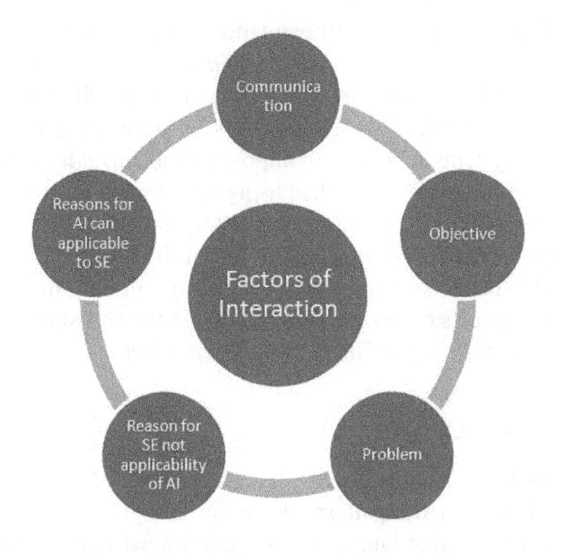

FIGURE 1.3 Factors of interaction between AI and SE.

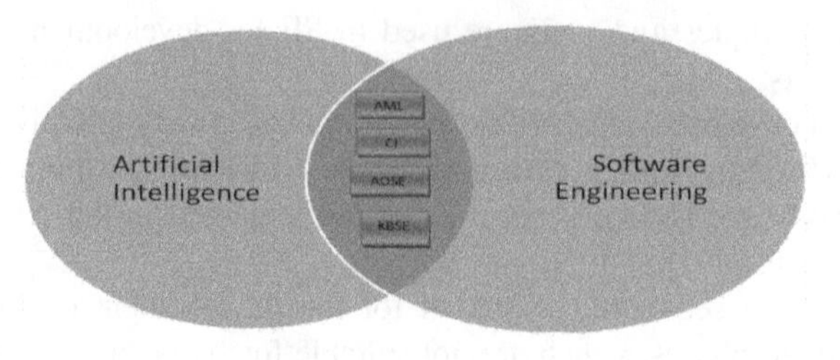

FIGURE 1.4 Research area of AI, SE, and intersection area between AI and SE.

Figure 1.4 represents the common research areas between AI and SE, which are as follows:

- **Ambient Intelligence (AmI)**: AmI is a promising discipline that adds intelligence to daily work and creates environments that are responsive to users. Its design and development depend on sensors, sensor networks, and AI. It is quickly establishing areas that help society through technology and allow people to be flexible and adaptive (Cook, Augusto, & Jakkula, 2009). Its main aim is to enhance the interaction between people and the environment. It makes the environment where we live and work more beneficial. The most common example of AmI is a smart system.
- **Computational Intelligence (CI)**: CI is the theory, design, application, and development of biological computational paradigms. Neural networks, evolutionary algorithms, and fuzzy systems are the three main pillars of CI (Aulakh, 2015). It is used for software security and reliability, to classify data, to assist in identifying and encrypting significant data, and to save time and processing manpower.
- **Agent-Oriented Software Engineering (AOSE)**: AOSE is an SE paradigm in which tiny intelligent systems help achieve the common objectives of both fields. These agents are a new area of AI and SE research intersection. From the AI side, this field focuses on developing more smart and self-directed systems to resolve composite problems where SE agents are used to design the software product using formal methods for development, verification, validation, and maintenance (Aulakh, 2015).
- **Knowledge-Based Software Engineering (KBSE)**: This is the application of knowledge-based systems technology in the design and production of software. It supports computer-aided design in the software design process and is applied in the entire life cycle of the product.

1.3 AI TECHNIQUES

a. **Case-Based Reasoning (CBR)**:

CBR is based on the pattern of human thought in cognitive psychology that concludes that human beings gain knowledge by solving several similar cases (Clifton & Frohnsdorff, 2001). CBR resolves a dilemma by recalling similar kinds of problems and assuming that they have a similar resolution. To solve the new problem, several past cases need to be solved to provide the solution or new methods. CBR recognizes that problems can be easily solved by repeated attempts (Chen, Jakeman, & Norto, 2008). CBR uses the following steps to solve a problem:

1. Retrieve the most appropriate past cases from the database.
2. Retrieved cases are used to find the solution of the new problem.
3. Revise the proposed solution of the new problem by several test executions.
4. After successful adaptation of the solution of the problem, save it for future use. (Figure 1.5).

Humans may generalize the patterns of cases into rules, and the solution of the problems can be provided by association with past similar case solutions. This is useful in fields where usually earlier rule-based reasoning is not able to solve the problem like machine learning, reasoning with incomplete information and knowledge acquisition.

b. **Artificial Neural Networks (ANN):**

ANNs work in the same way the human brain processes information. An ANN contains several processing units, called neurons or nodes, which work as a unit. Nodes are highly interconnected to other nodes of the same weight.

ANN nodes are arranged in several layers. The first layer is the input layer, whose work is to receive all the inputs. After receiving inputs, this layer processes the input data and returns outputs to the next layer as inputs. The

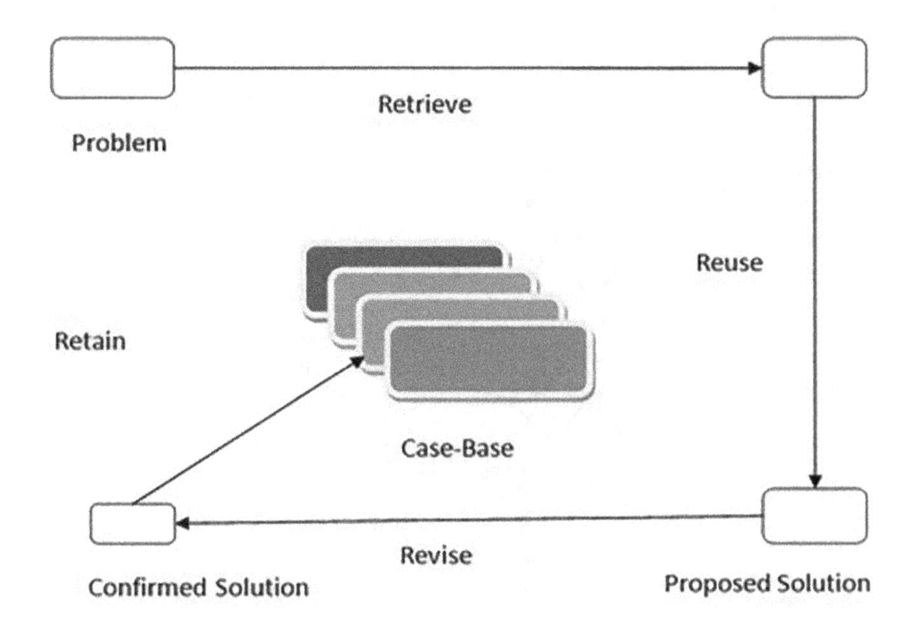

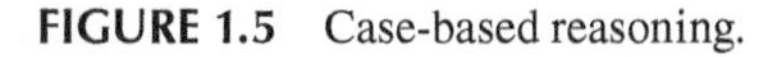
FIGURE 1.5 Case-based reasoning.

second layer is a hidden layer that is a collection of several layers of nodes with inputs and outputs. The final layer is the output layer. This layer receives inputs from the last layer of the hidden layer and converts received data into output that is useful to the user. Some earlier assumptions are used by ANNs, such as learning from examples by regulating the connection weightage of each and every node. In the ANN, learning can be categorized as supervised or unsupervised. Supervised ANN learning provides the exact or correct output to all input layers, while unsupervised learning provides different input patterns and then identifies the relationships among the given patterns and learns to categorize the input (Chen, Jakeman, & Norton, 2008) (Figure 1.6).

c. **Genetic Algorithm (GA)**:

A GA is a search technique that works until the satisfactory solution of the problem is not met with the filter solutions in a population surviving and passing their traits to offspring which replace the inferior solutions. Each potential solution is encoded, for example, as a binary string is called a chromosome and successive populations are called as the generation (Chen, Jakeman, & Norto, 2008). It is commonly used to generate high-quality solutions for optimization problems and search problems. GAs simulate the process of natural selection, which means the species that can adapt to environmental changes will survive and reproduce the next generation.

A GA consists of five phases:

1. Initial population
2. Fitness function
3. Selection
4. Crossover
5. Mutation

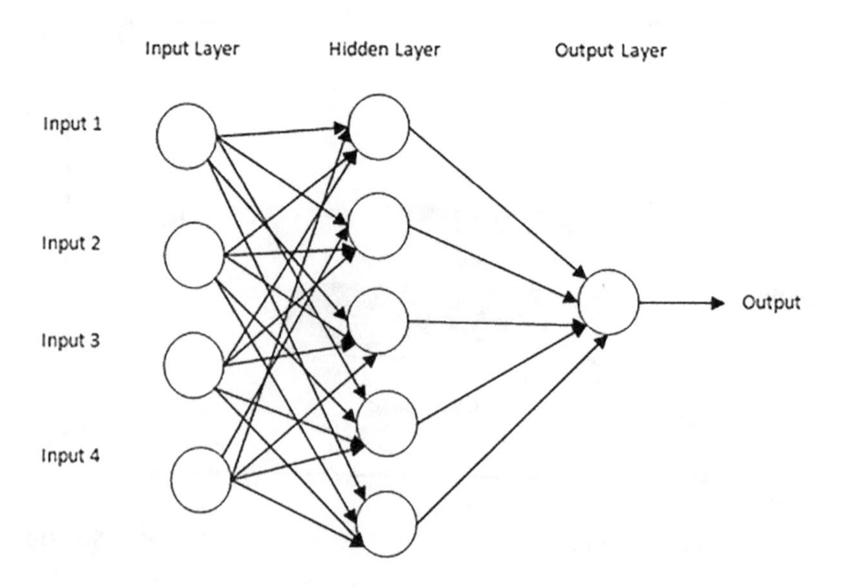

FIGURE 1.6 An example of an ANN.

The population of individuals is maintained within the search space, and each individual represents a solution of the given problem in the search space. Each individual is coded as a finite-length vector of components, and these variable components are similar to genes. The chromosome is a collection of several genes (Figure 1.7).

Once the initial generation is generated, the algorithm performs the following processes:

a. **Selection Operator**: This process selects the individuals who have good fitness scores and give them preference to pass their genes to successive generations of the individual.
b. **Crossover Operator**: In this process, mating of chromosomes is done between the randomly selected individuals. After mating, genes of individuals are exchanged and generate new individuals with different characteristics (Figure 1.8).
c. **Mutation Operator**: The main goal of mutation is to randomly add the genes in the generated individual to the crossover operator. For example (Figure 1.9):

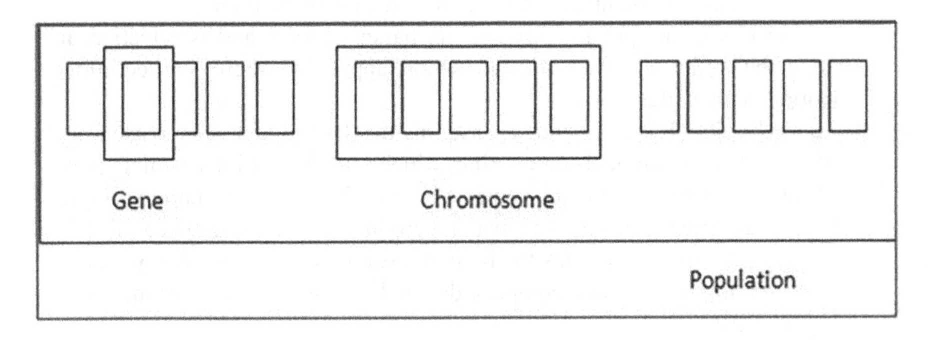

FIGURE 1.7 Chromosome.

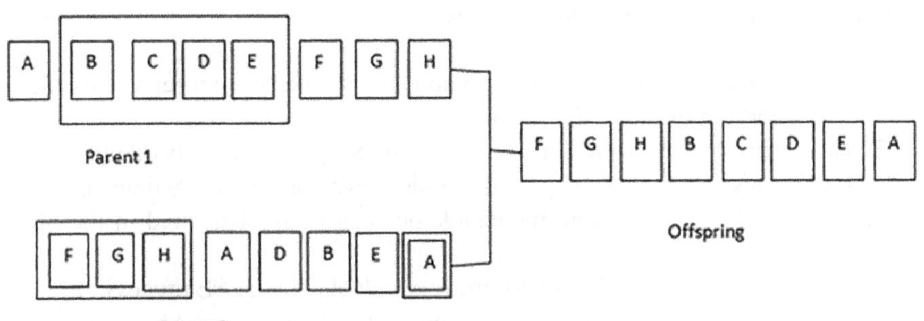

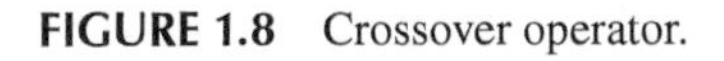

FIGURE 1.8 Crossover operator.

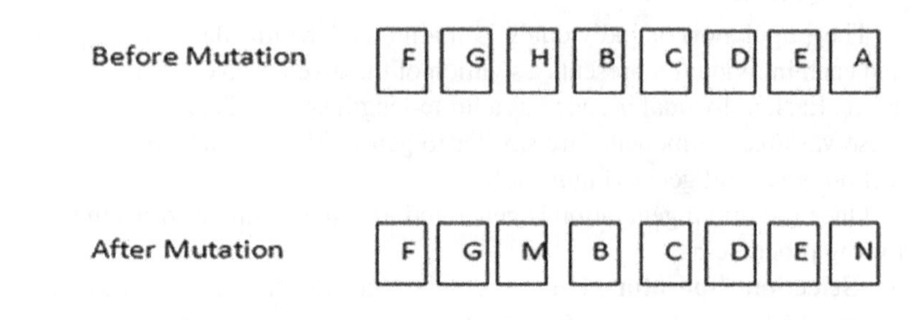

FIGURE 1.9 Mutation operator.

In summary, the GA performs the following tasks:

1. Arbitrarily initializes the population's as p.
2. Determines the fitness of the population.
3. Until the correct right result is obtained, the following steps are repeated:
    a. Select parents from the population.
    b. Crossover and generate a new population.
    c. Perform mutation on the new population.
    d. Compute the fitness of genes for the new population.

The GA is computationally simple, balances load, and is effective. It treats the model as a black box and is more applicable where detailed information is unavailable.

d. **Fuzzy Logic**: FL is a method of reasoning that looks like human reasoning. FL imitates human decision making, which includes all intermediate possibilities between the binary values YES and NO. The conventional logic block understands the precise input and produces a definite output as TRUE or FALSE, which represents the human YES or NO. FL uses fuzzy sets to deal with imprecise and incomplete data (Chen, Jakeman, & Norton, 2008; Figure 1.10).

1.4 WHY AI TECHNIQUES ARE IMPLEMENTED IN SE

The main reasons for implementing the AI method, tools, and techniques in SE are as follows (on the basis of our literature review):

1. Automatic programming (AP) is synonymous with SE and represents the new paradigm in future SE research.
2. The expert system is a major solution of the SE process and its dilemmas.
3. The SE environment of AI provides easier development and maintenance.
4. Several AI techniques and methodologies can be implemented in the SE field.
5. Rapid prototyping models of AI are very valuable in the SE process.
6. The cost of software products can be reduced with the help of AI techniques.
7. Error detection in coding will be isolated in the SE requirements phase.

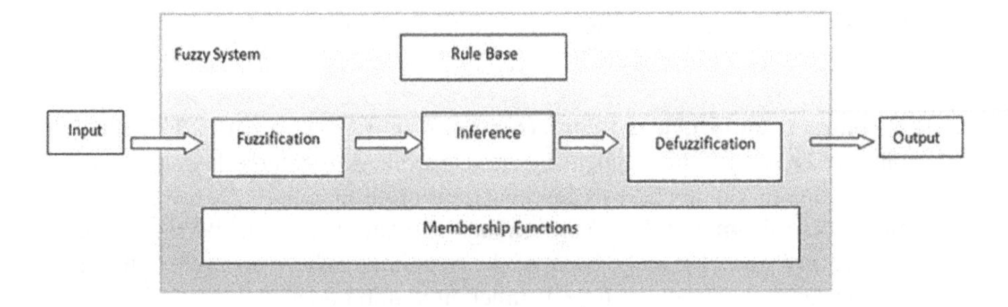

FIGURE 1.10 The main component of a fuzzy system.

1.5 IMPACT OF AI IN DIFFERENT PHASES OF SOFTWARE DEVELOPMENT

The development of software is becoming more complex these days because of the implementation of many functional and nonfunctional requirements. Sometimes the cost of the product is much more than the feasible cost of the product. The system and software architecture play vital roles in the development and maintenance of system software. The system architecture creates the software architecture and includes the system software requirements. The development of software starts with a requirements analysis, which produces the software requirement specification (SRS) document (Ammar, Abdelmoez, & Hamdi, 2012). The SRS includes the requirements for designing the software product and includes the functioning that is needed in the software. The software architecture design is followed by software coding in a specific programming language and testing of the software function before handing it to the customer.

AI is an important field of computer science whose aim is to create an intelligent machine. This field is mainly explained by the study and design of actions which are performed by human beings with intelligence in that situation. AI is basically focused on producing a machine that performs an action and is intelligent like a human being (Shankari & Thirumalaiselvi, 2014).

Applications of AI techniques in SE have grown tremendously. AI techniques are used in SE to minimize software product development time and increase software quality. That is why several AI techniques and tools are used by software engineers in software development (Ammar, Abdelmoez, & Hamdi, 2012).

1.5.1 Requirements Engineering (RE)

Requirements describe the "what" of a system, not the "how." The output of RE is one large document that is written in a natural language; that document contains a picture of the system and will indicate how that software product will perform. RE is the most critical activity of the system software process. The RE output is the requirements specification. The main functions of RE are gathering requirements, analysis of requirements, and reduction of the ambiguous representation of requirements.

The main problems faced during RE are as follows (Ammar, Abdelmoez, & Hamdi, 2012):

1. **Incomplete and Unclear Requirements**: Requirements are usually incomplete and unclear. This happens because customers do not really know what they want and are not able to describe their requirements.
2. **Some Requirements Are Ambiguous**: Software requirements should be written in natural language because the stakeholders have different cultural backgrounds. Natural language is inherently ambiguous.
3. **Some Requirements Are Conflicting**: This occurs when two different needs have the same resources or when the satisfaction of one needs to rule out another requirement.
4. **Some Time Requirements Are Volatile**: This means that requirements change from time to time. These changes may arise because the user's understanding of the functioning of a software product increases. If the requirements are not changed according to the need, then the requirements become incomplete and useless at that time.
5. **Requirements Are Not Easily Managed**: The main problem in RE is traceability. This means that it is difficult to trace the requirements from the point at which they were gathered to their implementation.

The main AI contributions in the requirement engineering phase are as follows (Meziane & Vadera, 2010):

- Develop tools that help recognize the requirements, which are written in natural language, and then minimize their ambiguous representation.
- In developing knowledge-based systems, AI should deal with the requirements.
- CI is used to classify requirements based on incompleteness and arrange the requirements based on their importance.

1.5.2 Software Architecture Design

The main challenge facing the developer in software architecture design is to design good software from the given requirements. The development of software architecture starts with defining the modules and their functions after understanding the requirements. Modularity, complexity, modifiability, understandability, and reusability of modules are the most common attributes that are used in architecture design (Ammar, Abdelmoez, & Hamdi, 2012). Modules can be defined as functioning units that perform a specific task. Modularity can be defined as the collection of manageable units or modules with well-defined interfaces among the modules. These modules are connected to each other with coupling and cohesion. The function of coupling is to measure the number of independent modules. When two modules have higher coupling, it can be said that these modules are strongly interconnected to each other and also depend on each other for their functioning. On the other hand, cohesion measures the dependency of elements of the same module. The designer should want to maximize the interaction within a module. Hence, the essential objective of software architecture design is to maximize module cohesion and minimize module coupling. In software architecture development, the GA technique is used to find the place in the software system from which decomposition can be implemented. To find and control data coupling, the AI fitness functions are used. Complexity and modularity of the software product are used to calculate the fitness function. The functioning of software architecture design is shown in Figure 1.11. The null architecture is used by the AI genetic algorithm to create the initial architecture. A fixed library of standard architecture solutions is employed to design new generations. The weightage of quality attributes is used to calculate the fitness function of the GA (Saini, 2016).

1.5.3 Risk Management (RM)

Developers of software products are extremely optimistic. When the developer plans to develop the software project, he or she presumes that the phases of software development will proceed according to plan. However, developers cannot exactly forecast everything because of the dynamic and inventive environment of software product development. This can lead to software surprises, and when an unexpected thing happens, move the project completely off track. These kinds of software surprises are not good at any time. RM became an important area in the software industry to reduce these surprise factors. RM is designed to solve a problem before it becomes a disaster. RM is the process of identifying and eliminating risk before it can damage the project. The RM process starts with the requirement study phase in SDLC and continues throughout the project development phase (Saini, 2016).

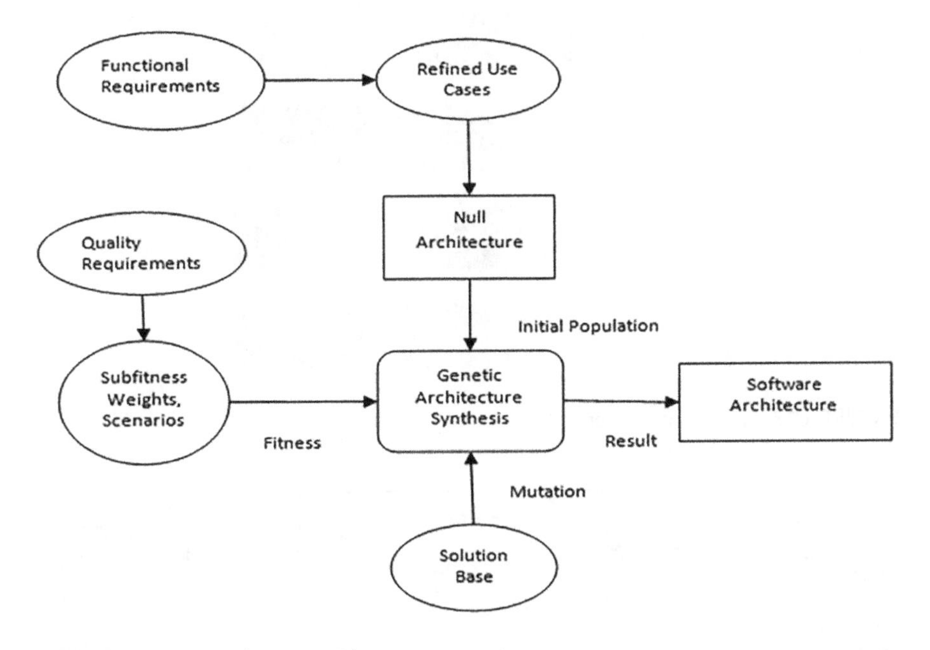

FIGURE 1.11 Evolutionary architecture generation.

RM involves the following steps, each of which is illustrated in Figure 1.12.

Risk Identification: In this step, software projects are investigated to identify areas where maximum risk can occur. This can be done to identify common software risk areas.

Risk Analysis: The main function of this step is to check the result after a software product modification in the input variables that may cause risk. It also identifies the area of the software project where the probability of risk is high and examines the impact of risk on the project. The main goal of this step is to find the risk factors.

Risk Remission: The function of risk remission is to manage the main risks that could cause a failure to reach the desired result. In this step, decisions are recorded, so customers and developers can take care to avoid or handle a problem when it arises. To fulfill this result, projects are observed early in the development phases to identify risks (Figure 1.12).

AI-based projects do not include RM strategies because automated programming techniques make the data structure flexible. The goal of automated programming is to make the specification less complicated, easier to write, easier to understand, and less prone to errors than the normal programming language (Raza, 2009) (Figure 1.13).

FIGURE 1.12 Risk Management (RM).

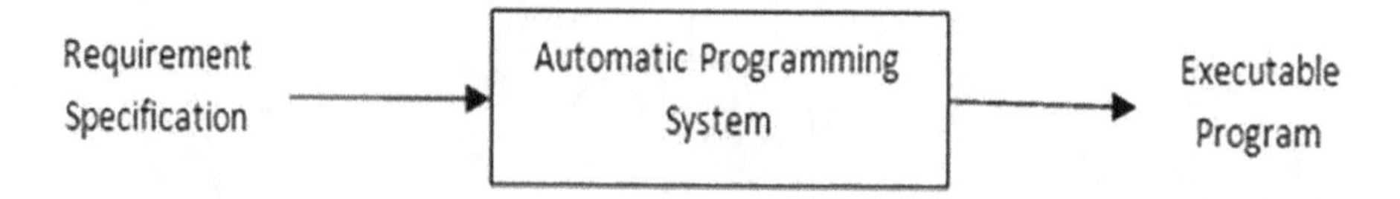

FIGURE 1.13 Automatic Programming System (APS).

1.5.4 Testing

Testing is the SDLC step that checks the software product. Software testing is an important field and requires significant effort. A proper strategy is required to carry out testing activities in a systematic and effective manner and provides a set of activities that are essential for the success of the project. The quality of the product depends on the effectiveness of the software testing. It also increases the number of satisfied users, lowers the maintenance cost of the product, makes the product more accurate, and produces a reliable result. In other words, an ineffective testing provides the opposite result – low-quality software, dissatisfied users, high maintenance costs, and unreliable and inaccurate results. That is why the software testing phase is a necessary and vital activity of the software development process. It is a very expensive process and represents almost one-third to on-half of the cost of a software product.

AI techniques have a special role software product testing. AI can generate test cases to check the behavior of the software product (Shankari & Thirumalaiselvi, 2014). AI tools are used to generate the test data, identify the suitability of test data, and analyze that coverage of test data on the software product. The AI techniques of GA, simulated annealing, and swarm intelligence are used to develop software product test data (Sorte, Joshi, & Jagtap, 2015).

1.6 TECHNIQUES OF AI

AI and SE are the main branches of computer science. One field is used to develop software in a systematic manner, and the other field makes the software more intelligent. AI has several techniques that are used in software development. Some AI techniques are explained in the following section (Shankari & Thirumalaiselvi, 2014) (Table 1.1).

1.6.1 Open Problems That Can Occur during the Application of AI Techniques to SE

The following problems can occur when applying AI techniques in SE:

1. **Software Development Approach**: The traditional method of project development uses phases like gathering requirements, analyzing requirements, designing software with a specific programming language, testing product functioning, and software deployment. The AI software development environment is relatively different. In an AI environment, data sources are classified, data is collected, and those data are cleaned. Each and every approach requires the special kind of skills and mindset of the developer.
2. **Quality of the System Depends on Data**: AI requires data to study the software product. AI depends on vast amounts of high-quality data to understand the behavior of the software product and quickly adopt the modifications that are required to improve the functioning of the product after

TABLE 1.1
Various AI Techniques and Their Purposes

AI Technique	Purpose
Knowledge-based system	Used in the design phase of the software development process
	Manages the requirement phase, planning, and project effort estimation
Neural network	Eliminates the risk associated with modules in software maintenance and used in the SE prediction outcomes
Fuzzy logic	Reasoning based on degrees of uncertainty
Genetic algorithm	Used in software testing and generating test cases
Case-based reasoning	Used for determining the duration or time taken to complete a project
Natural language processing	Helps in user requirements and improves the phase of the SDLC
Search Base Software Engineering	Reformulating SE problems as optimization problems.
Rule induction	Used to predict defects
Expert system	Uses knowledge to overcome RM strategies during the software development process
Genetic code	Develops automatically to generate computer programs and save time in the coding phase
Automated tool	Used for system redesign
	Changes the traditional software development to expert system development
Automatic programming	Generation of a program by a computer usually based on specifications
Simple decision making	Dealing with uncertainty
Intelligent agent	Generates a new intelligent software system for better communication
Probabilistic reasoning	Deals with uncertainty
Simulated annealing and Tabu search	Used in the field of engineering

data analysis. AI-based systems require a large amount of data to understand concepts or recognize features. Another important fact on which AI depends is the quality of the data that is used to design the product. The data should be of good quality and useful; otherwise, the AI-based product will not work properly.

3. **No Clear View**: AI-based projects do not have a clear view because of their experimental nature. Improvement of the project cannot be seen. This also makes it difficult to understand the whole concept for everyone.
4. **Building Trust**: AI works as a black box. Users don't feel comfortable using software when they don't know how the software works or how decisions are made.

5. **AI-Human Interface**: The increase of AI involvement in software design, development, and use, reduces the number of skilled persons who can work comfortably with this technology.
6. **More Initial Capital**: The costs of AI-based projects are very high, so every organization needs a large amount of initial capital to implement AI in the business.
7. **Software Malfunction**: With the implementation of AI in software, it is very complicated to recognize the reasons for software errors and malfunctions.
8. **High Expectation**: Developers and customers have high expectations for the functioning of AI-based software products, but the functioning of AI is not easily understood by them.

1.7 CONCLUSION

SE helps the developer design and develop a software product using SE principles but consumes very much time for the development. On the other hand, AI makes the software product more intelligent by using automated tools and techniques. Both fields – AI and SE – have their own advantages and disadvantages. To overcome the limitations of both fields, developers and researchers are striving to merge their advantages by implementing AI approaches in SE and vice versa. Therefore, the quality of the software product can be amplified by implementing the AI method in software development. By using automated AI tools or techniques, developers can abolish the risk, minimize software development time, and develop effective software products. In this chapter, we reviewed promising research work into finding how AI techniques help solve the problems faced by software engineers in the software product development period. We highlighted the impact of AI in software development activities like RE, software architecture design, coding, and the testing process.

The important outcomes of the above review are that when AI techniques are used in the SE process, they can minimize software development time, enhance product quality, and be used in the coding and testing phases. We also summarized some open problems in these research areas.

1.8 FUTURE SCOPE

The most important problem faced in the development of software products is transforming the requirements into a useful architecture, so that area of software product development needs to be improved using AI techniques. More work is needed on the coding and testing phases of software development using AI techniques to produce high-quality software products in less time and in an efficient way. The review shows that AI techniques are being used in the SE field, but more valuable study and effectiveness is needed in both fields to determine the effectiveness of different AI techniques. In the future, new research areas will include intelligent agents and application in the development of software.

REFERENCES

Ammar, H. H., Abdelmoez, W., & Hamdi, M. S. (2012). *Software Engineering Using Artificial Intelligence Techniques: Current State and Open Problems.* ICCIT, pp. 24–29.

Aulakh, R. K. (2015). A survey of artificial intelligence in software engineering. *International Journal for Research in Applied Science & Engineering*, 3 (4), 382–387.

Basu, T., Bhatia, A., Joseph, D., & L, Ramanathan. (2017). A survey on the role of artificial intelligence in software engineering. *International Journal of Innovative Research in Computer and Communication Engineering*, 5 (4), 7062–7066.

Chen, S. H., Jakeman, A. J., & Norton, J. P. (2008). Artificial intelligence techniques: An introduction to their use for modelling environmental systems. *Mathematics and Computers in Simulation*, 78 (2–3), 379–399.

Clifton, J. R., & Frohnsdorff, G. (2001). Applications of computers and information technology. In *Handbook of Analytical Techniques in Concrete Science and Technology.* Elsevier.

Cook, D. J., Augusto, J. C., & Jakkula, V. R. (2009). Ambient intelligence: Technologies, applications, and opportunities. In *Pervasive and Mobile Computing*, Elsevier, 277–298.

Harman, M. (2012). *The Role of Artificial Intelligence in Software Engineering.* RAISE. IEEE.

Jain, P. (2011). Interaction between software engineering and artificial intelligence: A review. *International Journal on Computer Science and Engineering*, 3 (12), 3774–3779.

Kumari, V., & Kulkarni, S. (2018). Use of artificial intelligence in software development life cycle requirements and its model. *International Research Journal of Engineering and Technology*, 5 (8), 1857–1860.

Meziane, F., & Vadera, S. (2010). *Artificial Intelligence Applications for Improved Software Engineering Development: New Prospects.* Information science reference, IGI Global, 278–299.

Raza, F. N. (2009, March). Artificial intelligence techniques in software engineering (AITSE). In *International MultiConference of Engineers and Computer Scientists (IMECS 2009)* (Vol. 1).

Saini, D. (2016). Applications of various artificial intelligence techniques in software engineering. *International Journal for Research in Emerging Science and Technology*, 3 (3), 25–33.

Shankari, K. H., & Thirumalaiselvi, R. (2014). A survey on using artificial intelligence techniques in the software development process. *International Journal of Engineering Research and Applications*, 4 (12), 24–33.

Sorte, B. W., Joshi, P. P., & Jagtap, V. (2015). Use of artificial intelligence in software development life cycle: A state of the art review. *International Journal of Advanced Engineering and Global Technology*, 3 (3), 398–403.

Tsai, W. T., Heisler, K. G., Volovik, D., & Zualkernan, I. A. (1988, August). A critical look at the relationship between AI and software engineering. In *[Proceedings] 1988 IEEE Workshop on Languages for Automation@ m_Symbiotic and Intelligent Robotics* (pp. 2–18). IEEE.

2 Software Effort Estimation

Machine Learning vs. Hybrid Algorithms

Wasiur Rhmann
Babasaheb Bhimrao Ambedkar University

CONTENTS

2.1 INTRODUCTION

Accurate estimation of effort is essential for proper allocation of resources during software development. Determining the efforts required for the development of the software is essential to estimate its cost. Although several techniques have been reported for effort estimation, it is still a challenging part of software development (Akhtar 2013). Software efforts calculated in person-month are helpful to estimate the cost of the software and project duration. Expert judgment, regression techniques, analogy-based effort estimation and machine learning techniques (MLTs) have been used by various researchers for software effort prediction. The performance of different techniques varies with changes in datasets. Although data from software projects are often noisy and incomplete, it is crucial for software effort prediction. The challenges associated with incomplete data and much of the effort estimation work done by researchers have focused on the construction of models (Sehra 2017). Cost estimation of software is affected by various factors, including types of programming languages, development methods and the experience of programmers.

In this chapter, machine learning and hybrid search-based algorithms (HSBAs) are applied for software effort prediction. HSBAs can be used for classification and regression problems. In our previous work, HSBAs were used to predict software defects (Rhmann 2018a, 2018b). Most of the available studies using HSBAs are based on classification problems. This work uses regression HSBAs for software effort estimation.

This chapter aims to address the following research questions:

RQ1: How do MLTs and HSBAs perform in terms of mean absolute error (MAE), root mean square error (RMSE), pred(0.30) and mean magnitude of relative error (MMRE)?

RQ2: Whether MLT and HSBA are statistically equivalent?

The major contributions of this chapter are as follows:

- Comparison of MLT and HSBA for software effort prediction
- Statistical assessment of the performances of MLT and HSBA for software effort estimation

This chapter is divided into eight sections. Section 2.1 is the introduction, Section 2.2 presents related work, Section 2.3 describes the dataset, Section 2.4 discusses the different techniques used for software effort prediction, Section 2.5 presents the metrics used to measure the performances of various techniques, Sections 2.6 and 2.7 present the results and statistical analysis and, finally, the conclusion is presented in Section 2.8.

2.2 RELATED WORK

Abdelali et al. (2019) reported an approach for software effort prediction using random forest. Three datasets obtained from software repository, namely, ISBSG R8, COCOMO and Tukutuku, were used for the experiment, and the results were evaluated using different performance measures, such as MMRE, median magnitude of relative error and Pred(0.25).

Zare et al. (2016) designed a model based on Bayesian network to predict software effort in person-month. Nature-inspired genetic algorithm and particle swarm optimizations were used to obtain optimal parameters. COCOMO NASA dataset was used for the experiment.

Fumin et al. (2017) presented an approach for software effort estimation using a novel AdaBoost and Classification and Regression Tree (ABCART). The presented model was evaluated on a dataset obtained from the GitHub repository.

Araujo et al. (2017) used MDELP, a hybrid multilayer perceptron-based model, for software effort estimation. The proposed model utilized the gradient-based learning technique, and the performance of the model was evaluated using MMRE and Pred(25).

Tirmula and Mall (2018) used a differential evolution algorithm for software effort prediction. They found their approach to be better than particle swarm-based and genetic algorithm-based algorithms.

Moosavi and Bardsiri (2017) presented an adaptive neuro-fuzzy inference system (ANFIS) and Satin bowerbird optimization (SBO) algorithm-based model for software development effort estimation. The proposed approaches were compared with biologically inspired algorithms.

2.3 DESCRIPTION OF DATASET

The datasets obtained from PROMISE software engineering repository Desharnais, COCOMO NASA v1 and COCOMO NASAv2 were used. These datasets contain information about 81, 60 and 93 software projects with 12, 17 and 24 attributes, respectively. Datasets were divided into 80:20 ratios for training and testing of data. All attributes of the Desharnais dataset are numerical, whereas some attributes of COCOMO NASA v1 and COCOMO NASAv2 are categorical variables. One-hot encoding was used to transform categorical attributes into numerical attributes to apply regression algorithms.

2.4 METHODOLOGY USED

The regression techniques MLT and HSBA were used to create models for software effort prediction. These techniques are described below.

2.4.1 Machine Learning

MLTs use past data of an event and then predict the occurrence of the same type of event in the future (Witten 2010). Machine learning-based regression models are applied for software effort estimation from historical data. In regression models, the target variable is continuous. CARET R package is used for applying machine learning algorithms (Kuhn 2008) and is described below.

2.4.1.1 Random Forest

Random forest combines the power of several decision trees and can work with categorical as well as numerical features. It is an ensemble technique that is useful for both classification and regression (Breiman 2001).

2.4.1.2 Artificial Neural Network

The artificial neural network is based on the concept of biological neurons in the brain and can be used for both classification and regression. These networks are also referred to as multilayer perceptron. In these networks, regression is performed by optimizing a function.

2.4.1.3 SVR (Support Vector Regression) (Linear)

In support vector machine for classification, a hyperplane is used to separate two classes. SVR is a nonparametric technique that relies on the kernel function for regression. The kernel function converts a nonseparable problem to a separable problem by transforming it into a higher dimension. This technique is less affected by outliers.

2.4.2 Hybrid Search-Based Algorithm (HSBA)

HSBAs are a hybrid of MLTs and search-based algorithms. The top three regression algorithms useful for software effort have been explored from the RKEEL package for experimental purposes (Moyano 2019). GFS_GP_R, GFS_GSP_R and FRSBM_R algorithms from the RKEEL package were used for the experiment.

2.5 METRICS USED TO MEASURE PERFORMANCE

Software efforts were predicted using MAE, RMSE, magnitude of relative error (MRE), MMRE and Pred(0.25). These are useful metrics for software effort estimation (Hamner et al. 2018) and are listed in Table 2.1.

2.6 RESULTS AND DISCUSSIONS

The performance results of MLTs obtained on three datasets are presented in Tables 2.2–2.4. The performance results of HSBAs in software effort prediction are presented in Tables 2.5–2.7.

TABLE 2.1
Metrics to Measure Performance

Metrics	Description	Formula
Mean Absolute Error (MAE)	Average of the absolute difference between the predicted value and observed value	$\text{MAE} = \frac{1}{n}\sum_{j=1}^{n} \lvert y_j - \bar{y}_j \rvert$
Root Mean Square Error (RMSE)	Standard deviation of the differences between predicted values and observed values	$\text{RMSE} = \sqrt{\frac{1}{n}\sum_{j=1}^{n} \lvert y_j - \bar{y}_j \rvert}$
Mean Magnitude of Relative Error (MMRE)	The mean of MRE for N projects is called MMRE	$\text{MRE} = \left\lvert \frac{\text{Actual effort} - \text{Predicted effort}}{\text{Actual effort}} \right\rvert$
Pred(l)	Percentage of projects with MRE values less than or equal to l	$\text{Pred}(l) = \frac{k}{N}$ *N* denotes the total number of observations and k denotes the number of observations with MRE<l

TABLE 2.2
Performances of Machine Learning Techniques on Desharnais Dataset for Software Effort Prediction

Technique	MAE	RMSE	Pred(0.30)	MMRE
Random Forest	2054.294	2846.61	0.312	0.6411
Artificial Neural Network	3011	3834.21	0.25	0.8902
Support Vector Regression (SVR)	2056.76	2946.09	0.437	0.302

TABLE 2.3
Performances of Machine Learning Techniques on COCOMO NASA v1 Dataset for Software Effort Prediction

Technique	MAE	RMSE	Pred(0.30)	MMRE
Random Forest	169.73	293.4579	0.5454	0.3691
Artificial Neural Network	688.50	1035.6864	0.090	4.247
Support Vector Regression (SVR) Linear	218.90	343.10	0.454	0.745

TABLE 2.4
Performances of Machine Learning Techniques on COCOMO NASA v2 Dataset for Software Effort Prediction

Technique	MAE	RMSE	Pred(0.30)	MMRE
Random Forest	416.142	682.11	0.312	0.995
Artificial Neural Network	656.44	931.67	0.12	9.195
Support Vector Regression (SVR)	517.76	917.79	0.187	1.618

TABLE 2.5
Performance of Hybrid Algorithms on Desharnais Dataset for Software Effort Prediction

Technique	MAE	RMSE	Pred(0.30)	MMRE
GFS_GP_R	5.058	6.55	0.176	0.778
GFS_GSP_R	6.117	7.745	0.352	1.66
FRSBMara_R	7.94	9.30	0.058	0.793

TABLE 2.6
Performance of Hybrid Algorithms on COCOMO NASA v1 Dataset for Software Effort Prediction

Technique	MAE	RMSE	Pred(0.30)	MMRE
GFS_GP_R	2.36363	3.2473	0.36363	0.6122
GFS_GSP_R	2.363636	3.765875	0.63636	1.123924
FRSBM_R	2.909	4.17786	0.36363	0.7384

TABLE 2.7
Performance of Hybrid Algorithms on COCOMO NASA v2 Dataset for Software Effort Prediction

Technique	MAE	RMSE	Pred(0.30)	MMRE
GFS_GP_R	6.5789	7.823	0.157	0.798
GFS_GSP_R	4.6315	5.99	0.368	1.013
FRSBM_R	8.263	9.621	0.105	0.764

It was observed that MAE and RMSE values of HSBAs were lower than MLTs, while pred(0.30) and MMRE values on different datasets were not uniform.

2.7 STATISTICAL ANALYSIS OF THE RESULT

Wilcoxon signed-rank test (Malhotra 2015), a nonparametric test, was applied to statistically assess the performances of HSBAs and MLTs for software effort prediction. The following hypotheses were formulated:

Null Hypothesis: The MLTs and HSBAs are equivalent in terms of their performance (RMSE) for software effort prediction
Alternate Hypothesis: The MLTs and HSBAs are not equivalent in terms of their performance (RMSE) for software effort prediction

```
Wilcoxon signed-rank test

data:  a and b
V = 45, p-value = 0.003906
          p-value<0.05
```

Hence, the null hypothesis is rejected, and MLTs and HSBAs are statistically different in terms of RMSE.

Null Hypothesis: The MLTs and HSBAs are equivalent in terms of their performance (pred(0.30)) for software effort prediction

Alternate Hypothesis: The MLTs and HSBAs are not equivalent in terms of their performance (pred(0.30)) for software effort prediction

```
wilcox.test(a,b, paired=TRUE)

 Wilcoxon signed-rank test

data:  a and b
V = 26, p-value = 0.7344
        p-value > 0.05
```

Hence, the null hypothesis is accepted. MLTs and HSBAs are statistically equivalent in terms of Pred(0.30).

2.8 CONCLUSIONS AND FUTURE SCOPE OF THE WORK

In the present work, software efforts were predicted with MLTs and HSBAs and statistical analysis was performed to assess the effectiveness of the different techniques in software effort prediction. For this, datasets obtained from software engineering repository were used. The results showed that HSBAs are better for software effort prediction than MLTs. In future research, datasets from software repositories like GIT can be extracted for software effort prediction.

REFERENCES

Abdelalia, Z., H. Mustaphaa and N. Abdelwahed. 2019. Investigating the use of random forest in software effort estimation. *Procedia Computer Science*. 148: 343–352.

Akhtar N. 2013. Perceptual evaluation for software project cost estimation using ant colony system. *International Journal of Computers and Application* 81(14): 23–30.

Araujo R.A., A.L.I Oliverira and S. Meira 2017. A class of hybrid multilayer perceptrons for software development effort estimation problems. Expert Systems with Applications 90(30): 1–12.

Breiman, L. 2001. Random forest. *Machine Learning* 45: 5–32.

Fumin Q., J.X. Yuan, Z. Xioke, X. Xiaoyuan, X. Baowen and Y. Shi. 2017. Software effort estimation based on open source projects: Case study of Github. *Information and Software Technology* 92: 145–157.

Hamner B. M. Frasco, and E.L. Dell. 2018. Evaluation Metrics for Machine Learning. https://cran.r-project.org/web/packages/Metrics/Metrics.pdf (accessed January 15, 2019).

Kuhn, M. 2008. Building predictive models in R using the caret package. *Journal of Statistical Software* 28(5): 1–26 https://cran.r-project.org/web/packages/caret/vignettes/caret.html, (accessed January 12, 2019).

Malhotra, R. 2015. *Empirical Research in Software Engineering: Concepts, Analysis, and Applications*. Boca Raton, FL: CRC Press.

Moosavi, S.H.S. and V.K. Bardsiri. 2017. Satin bowerbird optimizer: A new optimization algorithm to optimize ANFIS for software development effort estimation. *Engineering Applications of Artificial Intelligence* 60: 1–15.

Moyano, J.M., L. Sanchez, O. Sanchez and J. A. Fernandez. 2019. Using KEEL in R Code. https://cran.r-project.org/web/packages/RKEEL/RKEEL.pdf. (accessed January 12, 2019).

PROMISE Software engineering repository, http://promise.site.uottawa.ca/SERepository/datasets (accessed January 12, 2019).

Rhmann W. 2018a. Application of Hybrid search-based algorithms for software defect prediction. *International Journal of Modern Education and Computer Science* 4: 51–62.

Rhmann W. 2018b. Cross-project defect prediction using hybrid search based algorithms. *International Journal of Information Technology*: 1–8. DOI: 10.1007/s41870-018-0244-7.

Sehra S.K. and Y.S. Brara, N.K. Sukhjit and S. Sehra. 2017. Research patterns and trends in software effort estimation. *Information and Software Technology* 91: 1–21.

Tirmula, R.B. and R. Mall. 2018. DABE: Differential evolution in analogy-based software development effort estimation. *Swarm and Evolutionary Computation* 38: 158–172.

Witten I.H. and E. Frank, M.A. Hall. 2010. *Data Mining: Practical Machine Learning Tools and Techniques*. (The Morgan Kaufmann Series in Data Management Systems); 3 ed. Burlington, MA: Morgan Kaufmann (22 December 2010).

Zare, F., H.K. Zare and M.S. Fallahnezhad. 2016. Software effort estimation based on the optimal Bayesian belief network. *Applied Soft Computing* 49: 968–980.

3 Implementation of Data Mining Techniques for Software Development Effort Estimation

Deepti Gupta
Institute of Innovation in Technology and Management

Sushma Malik
Shobhit Institute of Engineering and Technology

CONTENTS

3.1 INTRODUCTION

Massive amounts of data in diverse formats are available in this binary world. This data is of no use until it is converted into meaningful and useful information. It becomes necessary to analyze this data and dig out relevant information from it. Thus, the field of data mining (DM) was introduced. DM is the practice of extracting meaningful information from large amounts of data. It is also referred to as the process of digging out knowledge in the form of unexplored and hidden patterns and relationships from data or knowledge discovery of data. It looks for hidden, valuable and potentially useful patterns from voluminous datasets that would allow businesses to take strategic decisions at the right time. DM is the act of automatically searching the huge amount of stored data to find the different trends and patterns that go beyond simple analysis methods. It utilizes complex mathematical algorithms for data segments and generates the probability of interesting patterns. It is all about discovering the unsuspected or previously unknown relationships between the datasets. Generally, DM is the process of analyzing huge amounts of data from different sources and shortening the data into useful and purposeful information (Sharma & Litoriya, 2012). This information can be used to boost revenue, cut down the cost of software product or both. DM software are analytical tools for analyzing the enormous amounts of data. These tools allow users to interpret the data from many distinct angles or dimensions, categorize it and outline the identified relationships between the datasets. Data collected from diverse sources are integrated into a single data storage. DM is a multistep process that includes accessing and preprocessing of data, implementation of the DM algorithm and analyzing the result and taking appropriate action on the basis of the result. Relevant data are retrieved from data storage, which are then preprocessed to make it consistent and desirable by applying different preprocessing techniques, such as removal of outliers and treatment of missing and null values. DM utilizes data analysis tools to find earlier unknown useful patterns and the relationship between the huge datasets. These tools can include statistical models, machine learning techniques and mathematical algorithms. Thus, DM analyzes data and performs pattern analysis. In several DM projects, certain techniques such as association, classification, clustering, prediction, sequential patterns and regression are developed and used. The discovered data can be stored in a data warehouse that is the central repository or the data marts that are the departmental datasets, which are then used by organizations to extract unknown patterns from preprocessed data to design business policies, take strategic decisions and solve various business problems. Hence, DM is nothing but finding out unexplored and substantial patterns and associations in the data to create business value. The various domains where DM has found its applications are market basket analysis, that is, figuring out the items in the basket of a customer at a store, generally used by banks and credit card companies to detect fraud, customer retention, suggest medicines to patients on the basis of symptoms, among other science exploration and military applications.

Software effort estimation is a process to predict the feasible cost of a product, a program or a project on the basis of available data. Software effort estimation has played a crucial role in the software industry and is a complex exercise in the domain of software engineering. Numerous estimation models such as software life cycle management (SLIM) by Putnam, constructive cost model (COCOMO) by Boehm and function points by Allan Albrecht have calibrated some predefined formulas for effort estimation on the basis of lines of code (LOC) and various other metrics. Correct and rigorous effort estimation in software development is influential for each kind of project in both overestimation and underestimation of cost in terms of development effort, as well as budget and time that can result in difficult situations for both the management of the organization and the software development team. On the one hand, if the project is underestimated, it can compromise the functionality and performance of the project due to lack of thorough testing; and on the other hand, overestimation can lead to overallocation of personnel and other development resources. Therefore, more consideration has been given to software estimation and its criticality has motivated research in this area. Thus, organizations are focusing on accurate estimations to be performed on historical data and are interested in predicting the development effort of the software at early stages so that the software development tasks can be planned and controlled, deadlines and budgets can be anticipated, resource allocation can be done, tasks can be scheduled and finally the directions can be given accordingly. To predict the effort of a project, software companies are using metrics from current and previous projects to identify the similarities between them. Software effort estimation has made project management and the allocation of resources required for a project much more easier.

3.2 LITERATURE REVIEW

Rajper & Shaikh (2016) identified various DM techniques used for cost estimation of software development. The software development environment is dynamic in nature, and factors of software development process are interrelated. Therefore, it's very difficult to find the most appropriate or suitable technique for cost estimation of software development. A hybrid cost estimation model has been proposed (Najadat, Alsmadi, & Shboul, 2012) to produce a realistic prediction model for software projects. DM techniques such as K-mean are used to classify the collected projects into clusters and then uncover the relationships between the projects. DM techniques are helpful in software engineering to handle bugs and manage unclear requirements that can affect the productivity and quality of software (Thomas, Joshi, Misal, & Manjula, 2015). DM techniques are also used for better effort estimation of software

(Sharma & Fotedar, 2014). Certain DM techniques are outlined by the author and are implemented on software engineering data to sort out challenges in software development, software debugging and maintenance of the software product. DM can also estimate the cost of the software development process. Software engineers and data miners should carefully employ the DM techniques to cut down the cost of tools, as well as gain insights into the available data (Lovedeep, 2014). Some software projects are not completed because of overbudgeting or failure in organizations. The main reason for this is inaccurate cost estimation during the early stages of software development. To solve this problem, a tool named 2CEE was developed for cost estimation of software using DM techniques. The accuracy of models has been validated internally through N-fold cross-validation (Ahmad, 2016). K-mean algorithm of DM is used to classify the project data into clusters on the basis of similarities. Code metrics is used to collect the data and extract metrics for a software project (Najadat, Alsmadi, & Shboul, 2012). DM includes a number of tools and techniques for analyzing the software data from different dimensions, as well as categorize and summarize the data on the basis of identified relationships. The DM tool Weka provides results on software cost estimation. The results show that the implementation of DM techniques into the existing software estimation techniques such as COCOMO can effectively improve the performance (Sharma & Litoriya, 2012). An overview of machine learning estimation techniques have been presented along with their advantages and disadvantages. The factors that can affect the estimation of a software project are the size of the project developing team, concurrency, fragmentation and complexity, as well as the computer platform where the software project can run. There are numerous methods to estimate the effort of the software but none of these methods can present the best estimates in all situations, and some techniques can be suitable for a single project but not for others. For example, case-based reasoning (CBR) performs well in helpdesk systems and artificial neural network techniques can be used to identify risk management and improve sales forecasting. Similarly, Genetic Algorithm (GA) technique is used in scientific research involving biological evaluation, whereas classification and regression tree (CART) technique may be useful in financial applications (Tripathi & Rai, 2017). Software development process includes a number of phases to design and develop the software, and each phase has its own importance and dependent on other phases. In each phase, complex and wide data are generated. DM techniques are used to extract the hidden data patterns and generate useful information, which is used to enhance the software development process by improving time constraints, budget of software and resources required, making it more reliable and maintainable (Thomas, Joshi, Misal, & Manjula, 2015).

3.3 DATA MINING

DM is the process of exploration and analysis of enormous amounts of data from different perspectives and summarizing the output into meaningful information (Deshmukh & Ahmad, 2016). Generally, the two forms of DM are descriptive mining and predictive mining. In descriptive mining, the data can be cauterized on the basis of data properties into the databases. Descriptive mining is used to extract meaningful data on past or recent events. However, predictive mining is used to predict future results on the basis of present behavior. Predictive mining deals with current data for prediction and provides answers for future queries (Ahmad & Bamnote, 2017). DM is also identified by other names such as information discover, knowledge extraction and data pattern processing (Smita, 2014). Although it is just one step in the knowledge discovery process, some researchers use it as a synonym for knowledge discovery process. In the process of discovering knowledge, all steps such as cleaning of data, integration of data collected from various sources, selection of required data, data transformation, DM and knowledge representation can be executed one after the other to obtain useful information from the database (Gupta & Devanand, 2013) (Figure 3.1).

Through DM relevant information is extracted from unstructured data. DM can be implemented with various techniques, and some of these are listed below.

3.3.1 Classification

Classification analyzes the data to extract and build models called classifiers on the basis of a training dataset that specifies important data classes and uses that model to predict or determine the categorical class labels for another dataset. Classification is a supervised learning algorithm because the output is known provided the input is available. For example, the students in a class can be grouped into three different categories – good, bad and average – in studies based on their percentage. Another example can be analyzing cancer data by a medical researcher to give the right treatment to a patient on the basis of symptoms. Other areas where classification finds

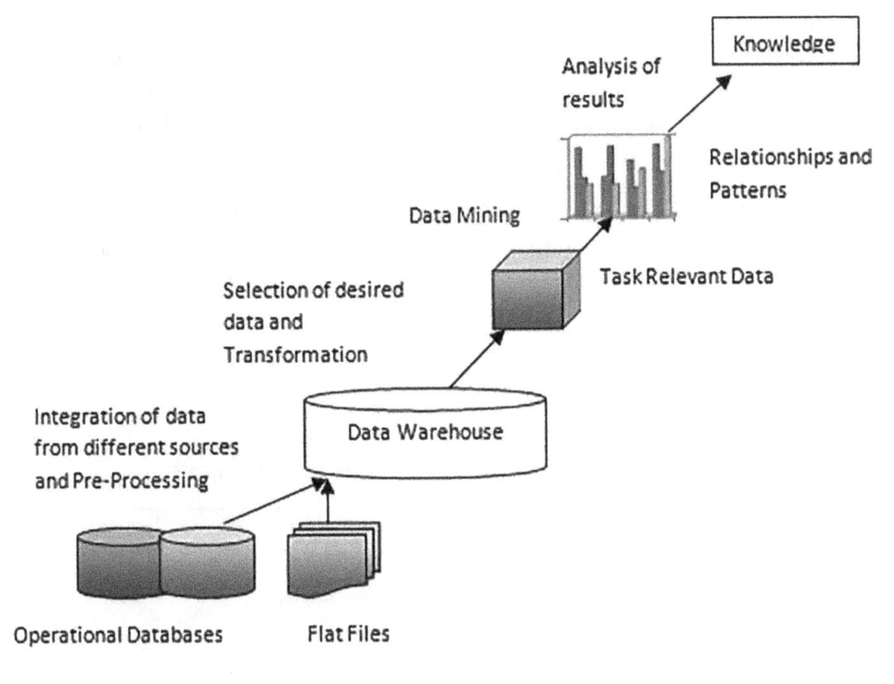

FIGURE 3.1 Knowledge discovery process.

applications include detection of frauds, predicting the performance of an individual or a company, diagnosing any disease or target marketing.

Data classification process includes two steps. The first step includes designing a classifier on the basis of previously known set of data classes or concepts. This is called the learning step (or training phase), where the technique is used to build and implement a classifier by training or "learning from" a sample training set consisting of input tuples and their related output class labels. The rows that are used for learning form the training set and are randomly sampled from the database under analysis.

The second step involves the use of the developed model for classification if the model has passed the accuracy test on the test set, after which the classifier can be used to identify the class of future data.

Various classification techniques are:

- **Decision Tree Induction**: It is the construction of decision trees from training tuples whose class labels are known in advance. A decision tree is in the form of a tree where attribute test is specified on each internal node (nonleaf node) and each branch shows the possible values for that particular attribute, and class labels are held by the leaf node. The topmost node of decision tree acts as the root node and can be decided by Gain ratio or Gini index. Rectangles in Figure 3.2 denote the internal nodes of a tree while the ovals represent the leaf nodes (Gupta & Devanand, 2013) (Figure 3.2).

  The algorithm to construct a decision tree is as follows. For each tuple of table for which class labels need to be determined, one branch of tree is traced in a top-down manner selecting that node, which satisfies the test condition specified on the path. Classification rules can be generated by performing an operation on various attribute conditions and values.
- **Rule-based Classification**: A set of if–then rules are extracted from the decision tree, which are then used for classification by reading the branch of the tree starting from the root moving downwards to each leaf node. As each node represents a certain attribute and all conditions are specified on the branches, the conditions are Anded to form a rule that denotes the

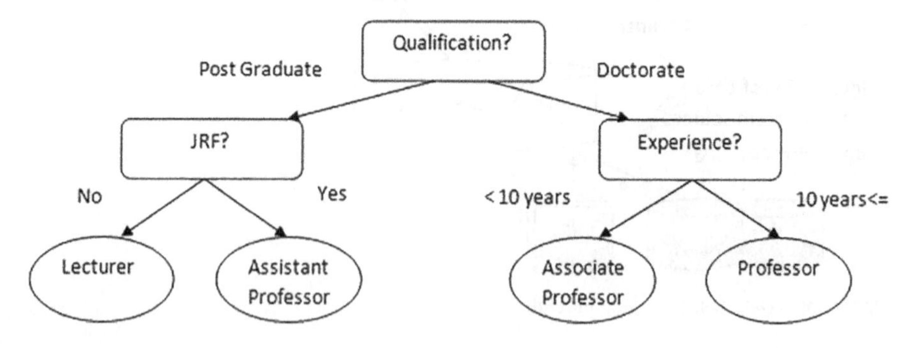

FIGURE 3.2 Decision tree induction.

If part of the statement, and the leaf node represents the Then part of the classification rule.

The general form of rule can be expressed as follows:

```
IF condition
THEN conclusion
```

For example, R1: IF qualification=doctorate AND experience=10 years THEN designation=professor

- **Backpropagation**: Backpropagation is another learning algorithm that uses a neural network consisting of several neurons connected through weighted connections. During the phase of learning the model, the network is trained by regulating the weights in multiple epochs or iterations feeding the network with a tuple in each epoch. The basic neural network consists of three layers: input, hidden and output. However, it can consist of multiple hidden layers depending on application requirements and the size of a dataset. After every iteration, the computed output is compared with the actual target value. The mean square error is then propagated back in the network to readjust the connection weights and nodes threshold. The process of backpropagation continues until the error, that is, the difference between the target and actual value, is minimized. Neural networks are used for numerical class labels or output (Figure 3.3).
- **K-Nearest Neighbor**: KNN classifiers work by analogy learning where an input tuple is compared to its similar rows in the data. Each tuple represents a point in an n-dimensional space. The similarity between the tuples is calculated using the distance metric, such as Euclidean distance. For KNN classification, the unknown tuple is assigned the most common class among its KNNs. When $k=1$, the unknown tuple is assigned the class of the training tuple that is the closest to it in pattern space. When given an unknown

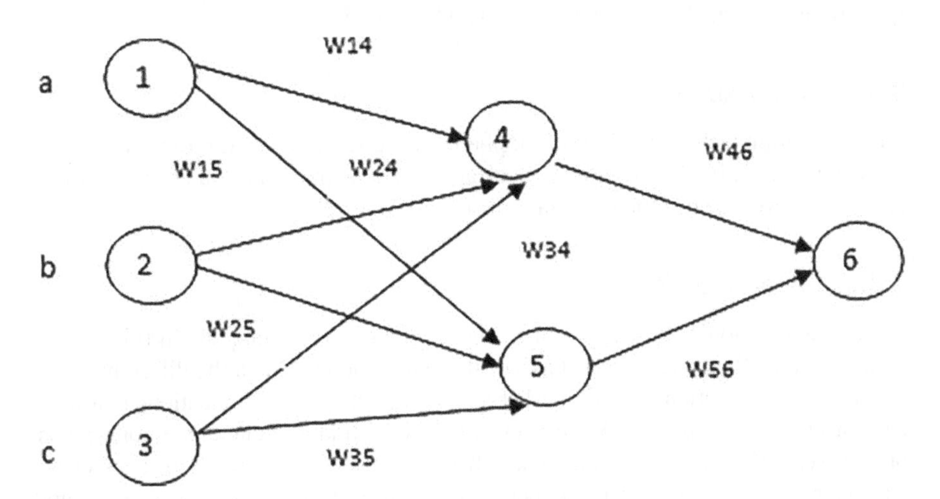

FIGURE 3.3 Backpropagation by neural network.

tuple, a KNN classifier searches the pattern space for the k training tuples that are closest to the unknown tuple. These *k* training tuples are the *k* "nearest neighbors" of the unknown tuple.

3.3.2 Regression

Regression is another DM technique that works on supervised learning. It helps to perform numerical prediction of the output provided a linear or quadratic equation is used for numerical input. The target value is known in the regression techniques. For example, the behavior of a child can be easily predicted using family history. It estimates the value by comparing already known value with predicted values (Gera & Goel, 2015). Regression is similar to classification, only with the difference that the predicted value is numerical not categorical.

3.3.3 Time-Series Analysis

Temporal data is analyzed by statistical techniques to perform time-series analysis. This technique is used to generate the future predictions (forecasts) based on already known past events (Kukasvadiya & Divecha, 2017). For instance, in stock market, the value of shares is predicted on the basis of past data (Smita, 2014).

3.3.4 Clustering

Unsupervised learning algorithm is used in clustering in which output of given inputs is not known. This technique forms the clusters of elements or tuples of tables on the basis of similar characteristics. This technique is unsupervised because the size of clusters and their forming tuples cannot be known in advance. For example, image processing and pattern recognition uses the clustering technique (Gera & Goel, 2015; Gupta & Devanand, 2013; Smita, 2014). The two main centroid-based partitioning algorithms to form clusters are k-means and k-medoid.

3.3.5 Summarization

Summarization technique of DM is applied to abstract the data. It provides an overview of the data. For illustration, a race can be precise in total minutes and seconds needed to cover the distance (Smita, 2014).

3.3.6 Association Rule

Mining association rules are one of the best known DM techniques. In this, a new relationship is discovered on the basis of the association between the different items in the same transaction. Market basket analysis is one of the applications that uses certain algorithms such as Apriori and FP-Tree to find associations among items on the basis of support and confidence that are likely to be purchased together by customers at any supermarket. Frequent item sets are generated at each level of algorithm, which satisfy the minimum support criteria; the association rules are then generated and filtered using the minimum confidence threshold (Gupta & Devanand, 2013) (Smita, 2014).

3.3.7 Sequence Discovery

Sequence discovery method is applied on database to uncover the relationship between the data. For example, it is used in scientific experiments, analysis of DNA sequence and natural disasters (Smita, 2014).

3.3.8 Text Mining

Text mining method is implemented on unstructured and semi-structured text data stored in word documents, email content, exchanged messages and hypertext files. Text mining incorporates document processing, its summarization, document indexing and mapping. It is normally used in the fields of education and business to analyze text data. Most organizations have large amounts of collected documents, and text mining can be implemented on that data to retrieve useful and interesting information (Madni, Anwar, & Shah, 2017). Text mining has also been applied in the field of sentimental analysis where the sentiments expressed by Internet users in the form of emojis, multilingual text, thumbs up and thumbs down are analyzed to find the opinion of people, whether positive, negative or neutral about a certain product, celebrity, movie or political figure.

3.4 SOFTWARE ENGINEERING

In computer science, software engineering is used to design and develop software programs or products. In software engineering, the development process is split into a number of phases to increase the output and improve both planning and management. Waterfall, prototype, iterative, incremental, spiral, rapid application development and Agile models are used to design and develop the software products. It's not necessary that one software development life cycle (SDLC) model is suitable for all software projects. An SDLC model is selected on the basis of project requirements, organizations and the developing team members (Sharmila & Thanamani, 2018). Software engineering for the development of software products heavily depends on four factors: people, product, process and projects. As software development is a people-centric activity, the success of the project in on the shoulders of the team members. In the software development process, a number of people are involved, such as developers, testers, project managers and users. However, for the success of software development, good managers are required. A good manager cannot ensure the success of the software projects but increases the probability of success. Therefore, selection of a manager for a software project is a crucial and critical activity. It is the responsibility of a manager to manage, motivate, encourage and guide people on the team. What the development team has to deliver to the customer? Obviously, the product: a solution to a customer problem. For the development of the product, the development team should clearly understand the requirements. Products can be structured or unstructured. In structured form, products may contain the source code, and in unstructured form, it may contain documentation and reports of bugs. The process is the way in which developing team produces the software. Software engineering includes a number of SDLC models and process improvement models. If the process is weak, then the end product will undoubtedly suffer. Although the process priority is after people and products, it plays a critical role in the success of the project (Xie, Thummalapenta, Lo, & Liu, 2009).

3.4.1 Software Engineering Data Sources Available for Mining

Software engineering comprises the various kinds of data that are used for analysis by DM. Numerous DM techniques can be implemented on the data generated by software engineering to find valuable knowledge. The following are the main sources of software engineering data (Thomas, Joshi, Misal, & Manjula, 2015).

3.4.1.1 Documentation

Data of software engineering documentation play a vital role in the development of software product, and this type of data is more complex. It mainly contains the data of software requirements as well as administrative data, written in natural languages. It also consists of the source code of the software written in some programming languages. Mining techniques can be used on this document to determine the valuable and unexplored information. Software documentation sometimes may also contain multimedia data in the form of audio or video instructions. Multimedia mining techniques are used to mine useful information from multimedia documentation.

3.4.1.2 Software Configuration Management (SCM) Data

In software engineering, SCM can be defined as the process that systematically manages, organizes and controls the modifications of data in documents and coding of the software product during the SDLC. Any modification in SCM documents influences the final product. It also stores bug tracking as well as revised software data. SCM maintains historical data of each revision of software as well as allows users and developers to access and revert to previous versions. Software metrics and the number of lines are analyzed with DM to extract interesting facts.

3.4.1.3 Source Code

Source code basically contains the instructions written by the programmer in human readable form while the programmer develops the software. A compiler is required to run the source code and change it into the machine code. To aid software maintenance and analyze software components, the source code of the software is utilized by DM. Source code becomes structured text after parsing. DM techniques are implemented on that text to predict the required changes after mining the change history, predict change propagation, as well as to predict the faults from cached history data.

3.4.1.4 Mailing Data

Mailing method is mainly used by users and developers to communicate with each other. Mail mostly contains large amounts of free text or data for analysis. It is easy to extract meaningful information or other graphs after analysis of mail data. However, the study of mail data is harder because all mails are interconnected with each other in the form of replies and always consider previous mails for meaningful extraction. Text analysis technique of DM is mainly used to analyze mail data.

3.4.2 Role of Data Mining in Improving the Development Process

DM has a crucial role in each and every phase of the software development process and improves the overall process (Figure 3.4).

3.5 SOFTWARE ESTIMATION

Project effort estimation of software is one of the significant topics in software development. Effort estimations are implemented during the planning stage of the software and help the software developer in controlling the work performance during the project developing phase, as well as help the software developer in decision making. Software estimation basically includes the project information, for example, how many people are required to complete the project and the duration of project development. Project estimation starts with the proposal of the project and continues throughout the project work. The estimation mainly includes the size estimation of project, efforts, schedule of project and overall cost estimation of the project. Accurate estimation of cost is very important for all projects as it may result in higher costs if not done correctly (Ahmad & Bamnote, 2017). Accurate cost estimation of project is critical for both customers and developers. Software projects use a number of models to calculate the cost of the project. However, it is difficult to say that there is any model that can provide cost estimation close to the actual cost of

Software Development Stage	Data Mining Techniques	Input Data	Data Analysis Result
Requirement Elicitation	Classification, Text Mining	Documentation	• Classify and prioritize the requirements. • Text mining technique is used to summarize the huge amount of data. • Saving time, cost and human resources.
Design	Clustering	Design document	• Clustering technique can differentiate the similar data in interval of times so the extraction and labeling of data becomes easier.
Implementation	Clustering, Classification, Text Mining and Association rules	Source code, SCM, Program Dependence graph	• Classification and text mining help the developer to recognize the possible bugs that might occur during integration of modules. • Bugs and failures can be detected easily and rectified. • Increasing the reliability and maintainability of the software.
Testing	Classification, Clustering	I/O variables of software system, Program executions	• Clustering and classification techniques can be used to find the bugs in a code of the software.

FIGURE 3.4 DM techniques in software engineering phases.

the project. Both overestimation and underestimation may occur due to inefficient resources, delay in the delivery of the final project and unexpected increase in project budget. Therefore, on that basis, no accurate decision can be made (Deshmukh & Ahmad, 2016). Effort estimation of the software is basically the predication of the efforts implemented during the designing and development of the software. The most important task of effort estimation is to find out the effort, cost and time, which is required to develop the software product. Hence, correct effort estimation helps the development team to manage time and budget throughout software development. SLIM and COCOMO are basic models that are used to calculate the efforts used in software development. If the developer has a good estimation at the planning phase, then the development of the software project can proceed smoothly (Sharma & Fotedar, 2014).

3.5.1 Traditional Techniques of Software Effort Estimation

Some of the traditional techniques of software cost estimation are discussed below.

3.5.1.1 Lines of Code (LOC)

LOC was the first traditional technique used for computing the effort estimation of a software project and it is easy to implement. LOC is a metric used to evaluate the software program according to its size. It cannot include comments or blank lines because their presence or absence does not affect the functions of the software. The quality of comments affects the maintenance cost because the maintenance of any software depends on the comments. However, too many comments also affect the maintenance efforts due to poor readability and understandability. Therefore, comments are not included while lines of program are counted for LOC. In LOC, productivity is normally measured in LOC/PM (lines of code/person-month). It is a simple method of counting but it is language-dependent. LOC of assembler is not the same as LOC of COBAL.

3.5.1.2 Function Point Analysis

FPA method of effort estimation in software engineering quantifies the size and complexity of system software in terms of functions. This method basically depends on the functionality of the program. It is independent of the language, tools and methodology used to design the software. The advantage of using FPA is that it helps in early effort estimation during software development. However, it is a time-consuming task because it works manually. Detailed knowledge of requirements is needed for the estimation, and new developers cannot easily estimate the size of the software with FPA (Tripathi & Rai, 2017). A system can be decomposed into functional units for FPA as follows:

- **Input:** Data or controlled information entering from outside the system.
- **Output:** Data or controlled information that is leaving or is sent outside the system.
- **Enquiries:** To retrieve data from the system that is made up of an input–output combination.

- **Internal Logical Files:** Logically related data maintained within the system.
- **External Interface Files:** Logically related data used by the system but maintained within another system.

3.5.1.3 Expert Judgment Method

In expert judgment method, a software estimation expert's or a group of experts' experience is used to calculate the efforts of new proposed projects. This method is used mostly to calculate software effort estimation. The requirements of past projects with proposed new projects can be easily differentiated by experts on the basis of their experience, and experts can also explain the impact of technique, architecture used to design software and language used for coding in the proposed project. However, at times, this method cannot give good results because it is not easy to document all factors used by experts, and experts may be biased (Tripathi & Rai, 2017).

3.5.1.4 Estimating by Analogy

In this method, the proposed project data is compared with the previously completed projects. To calculate the efforts of the projects, similar completed projects are identified from the stored database. It basically works on actual experience, and special experts are not required to calculate the efforts. However, for this method, large amounts of data from past projects are required, and sometimes similar projects are not handled previously, which results in prediction (Tripathi & Rai, 2017).

3.5.1.5 COCOMO (Constructive Cost Model)

COCOMO was proposed by Bohem in 1981. This model is used to predict the efforts and schedule on the basis of size of the software product. It is a regression model based on the number of LOC. COCOMO model is categorized into three types as follows:

1. **Organic:** A software project can be called organic when the development team size is small, the problem is well understood by the developing team and the team members are nominally experienced in designing and developing similar projects. Simple business systems like payroll and data processing systems are some examples.
2. **Semidetached:** A software project can be called semi-detached type when the development team is basically a mixture of both experienced and inexperienced team members. The developing team members have finite experience in developing related systems but have not tried some aspects of the software being developed. At that time, team members need some more experience and guidance for designing and developing the software product. Designing and developing of compiler, editor and database management systems are some examples of this type.
3. **Embedded:** A software project can be called embedded type when the developed software is strongly coupled with the hardware of the system. The developed system has a high level of complexity and experienced team members are required to complete the software project. Air traffic control and ATM are some examples of embedded model (Figure 3.5).

Model	Size of Project	Project Nature	Deadline of the Project
Organic	2-50 KLOC	• Small Size • Experienced Developer • Familiar Environment • Example: Pay roll	Not Tight
Semi Detached	50-300KLOC	• Medium Size • Medium size Team • Average previous experience developer • Example: Database System, Editors	Medium
Embedded	Over 300KLOC	• Large Project • Real time systems • Complex Interface • Very little previous Experience • Example: ATM, Air Traffic Control	Tight

FIGURE 3.5 Comparison of COCOMO models.

Project	a_b	b_b	c_b	d_b
Organic	2.4	1.05	2.5	0.38
Semidetached	3.0	1.12	2.5	0.35
Embedded	3.6	1.20	2.5	0.32

FIGURE 3.6 Basic COCOMO coefficient.

The basic COCOMO equations are:

$$E = a_b\,(\text{KLOC})\,b_b$$

$$D = c_b\,(E)\,d_b$$

Where E=Efforts applied in person-months

D=Development time in months

The coefficients a_b, b_b, c_b and d_b are constants and are shown in Figure 3.6.

When effort and development time of a software project are known, the required average staff size to complete the project and productivity can be calculated as:

$$\text{Average staff Size (SS)} = \text{E/D Persons}$$

$$\text{Productivity (P)} = \text{KLOC/E}$$

3.5.2 Data Mining Techniques of Software Effort Estimation

DM extracts meaningful and useful patterns or information from a large amount of data. It is a powerful technique and a novel technology that helps software development companies to extract and focus on the most significant information in their data warehouse. Software effort estimation plays an important role in software engineering. Estimation of efforts is done during the planning or requirement gathering phase. DM techniques are used to improve the estimation of software. Effort estimation basically predicts the time and cost required during the development of the software product. In recent years, effort estimation has become an essential and most analyzed variable before starting to develop a new software product. Figure 3.7 shows some DM techniques used in software estimation. Some of them are explained below.

3.5.2.1 K-Mean

The K-mean technique can partition the datasets in K clusters, and K is a positive range. This DM technique partitions the datasets into predefined and nonoverlapping clusters where each data should belong to only one cluster. The grouping is completed by minimizing the total of squares of distances between centroid and data. Thus, the main aim of K-mean is to classify the data into specific clusters (Sharma & Fotedar, 2014) (Figure 3.8).

3.5.2.2 KNN (K Nearest Neighbors)

KNN is a supervised classification DM technique . It uses a group of labeled points and then uses them to label other points. For labeling a new point, the closest labeled points to that new point are searched. "k" in KNN is the number of neighbors it

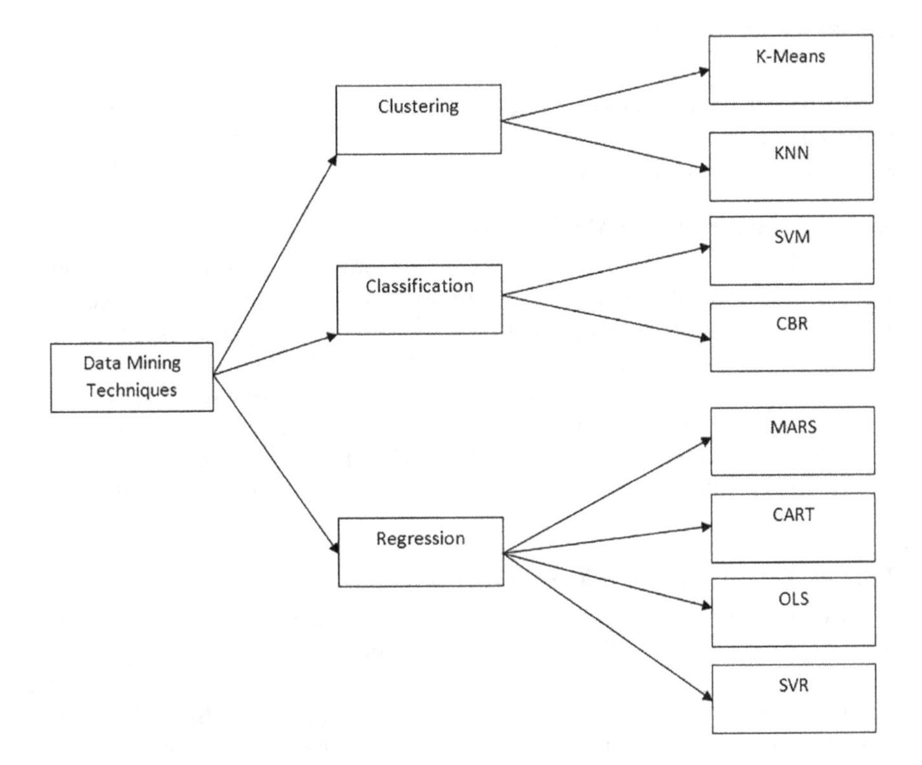

FIGURE 3.7 DM techniques.

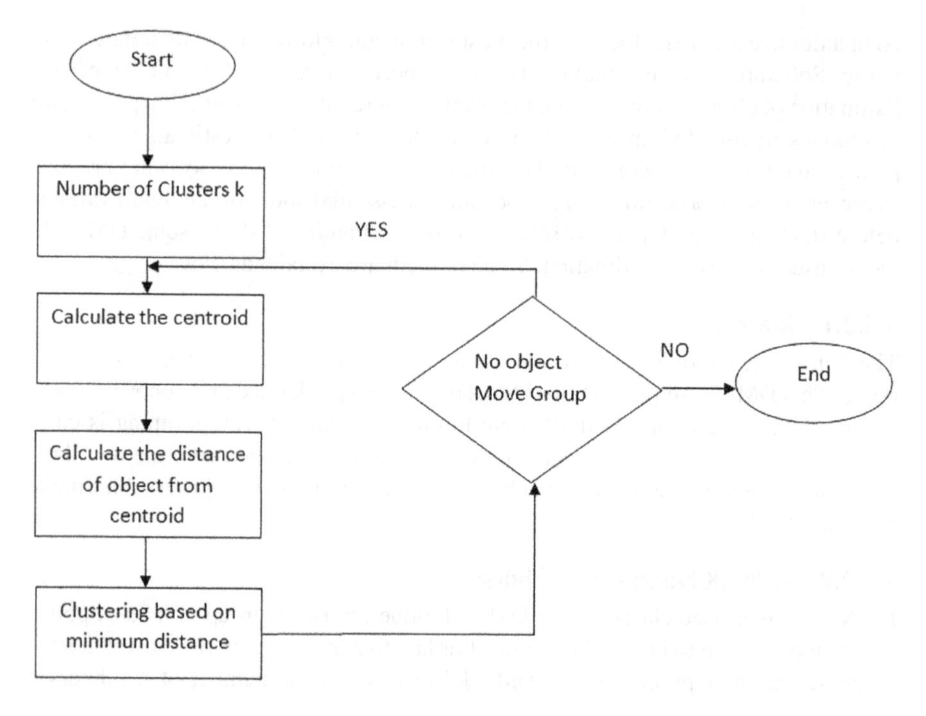

FIGURE 3.8 K-mean technique.

checks. This is called a supervised technique because it attempts to classify a point based on the known classification of other points (Khalifelu & Gharehchopogh, 2012; Sharma & Fotedar, 2014).

3.5.2.3 Support Vector Machine (SVM)

SVM is the most used classifier of machine learning. It is a supervised learning algorithm that uses the concept of margin to classify data into classes. The main aim of SVM is to design the best boundaries for the decision into classes and easily identify the class of input data in the future. That boundary of decision is known as a hyperplane. In this, data is partitioned into training and testing sets. Each instance of data in the training set has one value called class label, as well as enclosed number of attributes called observed variables. Its combination with k-means can be used to improve the efficiency of effort estimation in software engineering (Sharma & Fotedar, 2014).

3.5.2.4 CBR (Case Based Reasoning)

In CBR, previous experience is used to understand and resolve a new problem (Clifton & Frohnsdorff, 2001). A problem solver remembers preceding circumstances which are similar to the present situation and uses their experience to resolve the current problem. It is used for classification and regression. This technique can be implemented with partial knowledge of the target project. It is a simple and flexible technique (Tripathi & Rai, 2017).

3.5.2.5 MARS (Multivariate Adaptive Regression Splines)

MARS is a new nonlinear and statistic regression technique of DM introduced by Milton Friedman. Its main objective is to forecast the continuous dependent variable value or result variable on the basis of autonomous or predictor variable. This technique does not impose any relationship between the predictor variables and the result variables, instead drives useful models between them (Dejaeger, Verbeke, Martens, & Baesens, 2012).

3.5.2.6 CART (Classification and Regression Tree)

The CART technique of DM was developed by Leo Breiman, Jerome Friedman, Richard Olshen and Charles Stone in 1984. It is a nonparametric technique and easily handles the outliers. CART is used to generate outputs. With the help of classification tree, the class of target variable is identified, and regression tree is used to predict the value of the target variable (Sharma & Fotedar, 2014). In CART, variables are not selected in advance because this technique itself recognizes the variables based on their importance and eliminates the nonimportant variables (Tripathi & Rai, 2017).

S.No.	Software Effort Estimation Techniques			
	Technique	Key Idea	Advantage	Disadvantage
1	LOC	Evaluate the software program according to its size	Easy to implement	Language dependent
2	FPA	Quantified the size and complexity of system software in terms of functions	independent of the language, tools and methodology which are used to design the software	Detailed knowledge of requirements is needed for the estimation and new developers are not easily estimate the size of software with function point analysis
3	Expert Judgment	Based on the judgment of experience of the experts	Simple to understand	Helpful only if new Software project is similar to earlier software
4	Estimating by Analogy	Based on previously completed projects	Not special experts are required	Lots of data of past projects are required
5	COCOMO	Effort and cost are predicted based on the size of the software	Easily adjusted according to needs of the organization.	Should have proper knowledge about the size of the project.
6	K-Mean	Based on clustering of data by distance between centroid and data.	Very fast computation and simple to understand.	It is applicable only when mean is defined. Number of clusters should be known in advance.
7	KNN	Use voting of neighbors	It is simple so used for recognition problems well.	It is a lazy learner as there is no need to train the data.
8	SVM	A non-linear machine Learning technique based on classification and regression.	SVM is less over fitting and optimally separate the data	There is problem of choosing kernels. There are discrete data obtained by which more problems can be created.
9	CBR	Analogy cases are made and used.	It is easily understandable and Useful where domain is difficult to model. Potential to lessen the problem of outliers.	A complex computation is required
10	MARS	Works on nonlinear relationships	It is used for capturing Communication between variables	It has low dimensionality because of nonparametric smoothers.
11	CART	Tree based approach	It is simple to use and can easily handle complex situations.	Unstable samples
12	OLS	Based on fits linear Regression function.	It is simple method and easy to understand.	An attribute is removed, if more than 25% of the attribute values are missing. It Cannot handle highly correlated values.

FIGURE 3.9 DM techniques for software effort estimation (n.d.).

3.5.2.7 OLS (Ordinary Least Square)

OLS is the oldest and most used technique of DM, which is used to calculate software system effort estimation. OLS is the best estimator for statistics in an unbiased manner. This technique minimizes the sum of square of the difference between an observed variable and a predicted variable (Sharma & Fotedar, 2014) (Figure 3.9).

3.6 CONCLUSION

We studied two different domains of computer science: software engineering and DM. With the help of integration of DM techniques and software cost estimation models, we classified the project and generated approximate estimation of a new project based on data from past projects. We have shown the need and importance of DM techniques /in software engineering to estimate the efforts in the development of a software product. On implementing DM techniques, the accuracy of software efforts can be improved. In this chapter, we elaborated on some DM techniques such as K-mean, KNN, OLS, CART, MARS, and CBR to improve the accuracy of software effort estimation.

REFERENCES

Ahmad, S. W. (2016). The effects of classification data mining algorithms on software cost estimation. *International Journal of Innovative and Emerging Research in Engineering, 3*(1), 218–221.

Ahmad, S. W., & Bamnote, G. R. (2017). Software cost estimation using data mining. *International Journal of Scientific Development and Research (IJSDR), 2*(7), 181–183.

Clifton, J. R., & Frohnsdorff, G. (2001). Case based reasoning. *Handbook of Analytical Techniques in Concrete Science and Technology.* eBook ISBN: 9780815517382.

Dejaeger, K., Verbeke, W., Martens, D., & Baesens, B. (2012). Data mining techniques for software effort estimation: A comparative study. *IEEE Transactions on Software Engineering, 38*(2), 375–397.

Deshmukh, S. A., & Ahmad, S. W. (2016). Using classification data mining techniques for software cost estimation. *International Journal of Scientific & Engineering Research, 7*(2), 276–278.

Gera, M., & Goel, S. (2015). Data mining - techniques, methods and algorithms: A review on tools and their validity. *International Journal of Computer Applications, 113*(18), 22–29.

Gupta, V., & Devanand, P. (2013). A survey on data mining: Tools, techniques. *International Journal of Scientific & Engineering Research, 4*(3), 1–14.

Khalifelu, Z. A., & Gharehchopogh, F. S. (2012). Comparison and evaluation of data mining techniques with algorithmic models in software cost estimation. *Procedia Technology, Elsevier*, 65–71.

Kukasvadiya, M. S., & Divecha, N. H. (2017). Analysis of data using data mining tool. *International Journal of Engineering Development and Research, 5*(2), 1836–1840.

Lovedeep, V. K. A. (2014). Applications of data mining techniques in software engineering. *International Journal of Electrical, Electronics and Computer Systems, 2*(5), 70–74.

Madni, H. A., Anwar, Z., & Shah, M. A. (2017). Data mining techniques and applications – A ResearchGate. doi:10.23919/IConAC.2017.8082090. IEEE.

Najadat, H., Alsmadi, I., & Shboul, Y. (2012). Predicting software projects cost estimation based on mining historical data. *International Scholarly Research Network.* doi:10.5402/2012/823437.

Rajper, S., & Shaikh, Z. A. (2016). Software development cost estimation: A survey. *Indian Journal of Science and Technology, 9*(31). doi:10.17485/ijst/2016/v9i31/93058.

Sharma, M., & Fotedar, N. (2014). Software effort estimation with data mining techniques- A review. *International Journal of Engineering Sciences & Research Technology, 3*(3), 1646–1653.

Sharma, N., & Litoriya, R. (2012). Incorporating data mining techniques on software cost. *International Journal of Emerging Technology and Advanced Engineering, 2*(3), 301–309.

Sharmila, S., & Thanamani, A. S. (2018). Analytical study of data mining techniques for software testing. *International Journal of Analytical and Experimental Modal Analysis, 10*(12), 31–40.

Smita, P. S. (2014). Use of data mining in various field: A survey paper. *IOSR Journal of Computer Engineering, 16*(3), 18–21.

Thomas, N., Joshi, A., Misal, R., & Manjula, R. (2015). Data mining techniques used in software engineering: A survey. *International Journal of Computer Sciences and Engineering, 4*(3), 28–34.

Tripathi, R., & Rai, P. K. (2017). Machine learning method of effort estimation and it's performance evaluation criteria. *International Journal of Computer Science and Mobile Computing, 6*(1), 61–67.

Xie, T., Thummalapenta, S., Lo, D., & Liu, C. (2009). Data mining for software engineering. *Institutional Knowledge at Singapore Management University*, 55–62. doi: 10.1109/MC.2009.256.

4 Empirical Software Measurements with Machine Learning

Somya Goyal
Manipal University Jaipur, Jaipur, Rajasthan
Guru Jambheshwar University of
Science & Technology, Hisar, Haryana

Pradeep Kumar Bhatia
Guru Jambheshwar University of
Science & Technology, Hisar, Haryana

CONTENTS

4.1 INTRODUCTION

Measurement of the attributes of software processes, products, projects and people associated with the software development is necessary so that the industry can deliver quality product, that is, high-quality software within the limits of time and cost. It is evident that accurate software measurements using empirical techniques are essential. As per the Chaos Report (Chaos Report 2015), only 23% of total projects get the status of "successful project completion." The reason for this poor successful completion rate is the inaccurate measurement of attributes of software quality and quantity (Demarco 1982). Empirical techniques are essential for accurate measurements in the field of software engineering. We need to evaluate, assess, predict, monitor and control the various aspects of software development. For successful project completion, the quantitative methods need to be followed. This chapter discusses the empirical approach for software measurements using machine learning (ML) techniques. A majority of research work has already been done in this field; ML has found software measurements a very fertile ground. Both dimensions including software quality and quantity can easily be measured empirically using ML techniques. Software quantity measurement is analogous to effort estimation, cost estimation, schedule prediction and several software measurement tasks, which can be modeled as regression-based tasks. Software quality measurements is analogous to defect prediction, quality prediction, prediction of faulty modules and other such problems, which can be formulated as classification tasks in the world of ML. In this way, software quantity and quality measurements together can be formulated as supervised ML-based problems. This is the base point which is being utilized in this research field for measuring software using ML techniques empirically. Since the 1980s, this field is resonating with software researchers, which is quite fascinating. This chapter demonstrates the usage of ML techniques for both software quality and quantity measurements. With a basic introduction to the current trends of the field and moving through problem definition, we will reach the experimental set-up and then draw inferences from the experiments. This chapter aims to provide the reader practical and applicable knowledge of ML and deep learning for empirical software measurements.

4.2 MACHINE LEARNING TECHNIQUES FOR EMPIRICAL SOFTWARE MEASUREMENTS

The successful development and delivery of high-quality software is highly desirable for satisfying the specified requirement constraints. All the current market trends lead to the excessive use of ML for analyzing, assessing, predicting and measuring the various quality and quantity attributes of ongoing development projects at a small and/or large scale. ML techniques are being deployed for estimating the cost of entire projects and predicting the quality of the software in the early phases of project development, so that accurate early predictions can be used later for strategic decision-making, and ultimately, increasing the success rate of software development projects. Accurate predictions made in the early development phases have immense impact on the probability of successfully delivering high-quality projects on time and at the estimated cost. Software measurements are empirically made in two dimensions—software quality and software quantity. To deploy ML techniques for empirical software measurements, the first step is to formulate the software measurement problems as learning problems. Once this is done, the next step is to know the ML techniques available at our disposal and select the suitable ones for our software measurement problems. These two crucial steps are discussed in the following subsections of this chapter. Then, we take a look into the current market trends in the field of software industry projects, which are intensively using ML as a predictor for their projects.

4.2.1 Formulating "Measuring the Software" as a "Learning Problem"

The very first step in empirical software measurement using ML techniques is to formulate software measurement problems as learning problems. Measuring the software empirically is done in two aspects, namely, measuring the attributes of software quality and measuring the software quantity. Measuring the software quality empirically refers to predicting the probability of the modules of the software being fault-prone in very early development stages that can endanger the successful delivery of the entire software project in the final development stages. This refers to the prediction of whether a particular module of the project would be highly risk-prone or low risk-prone. This aspect of empirical software measurements is also referred to as fault prediction. This aspect of software measurement can be formulated as classification type—supervised ML problem. Measuring the software quantity empirically refers to estimating the efforts or cost incurred in software development. This aspect of empirical software measurement is also known as cost prediction or effort prediction. Effort prediction can be formulated as regression type—supervised ML problem.

Software engineering has emerged as a field which is a fertile ground for ML algorithms. Several software engineering tasks like various development and maintenance tasks that come under software engineering can be formulated as learning problems and can be handled as application of learning algorithms. ML is a subfield of artificial intelligence, as shown in Figure 4.1.

4.2.1.1 Quality Prediction as Classification Problem

In empirical software measurement, quality prediction is a major aspect. The quality of a software is directly associated with the fault-proneness of the software modules. A fault is an error during any development phase, which if not identified and rectified at early stages can result in the failure of the entire project. Therefore, it is essential for the project development team to predict the risk of a module being faulty. This

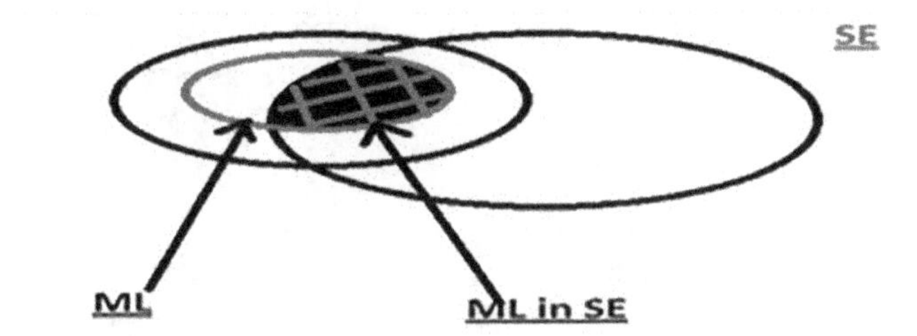

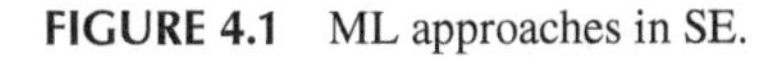
FIGURE 4.1 ML approaches in SE.

is to ensure that testing efforts can be focused on that particular module, and hence increasing the chances of not resulting in failure caused by that fault.

To predict the fault-proneness, relevant attributes of the project, such as cyclomatic complexity or length of code (LOC) measures, can be considered from the past data. Subsequently, the risk of a new project can be predicted to make strategic decisions for the testing phase and other phases of the software development project.

Fault-proneness of a module can be modeled as a two-class classification problem, where high-risk module and low-risk module are the two classes of classification problem. The information about the module to be classified makes up the input for the ML-based classifier, and the output will be {0,1}, denoting one of the two classes (Class 0=not so risky; Class 1=risky), as shown in Figure 4.2. In the figure, on x-axis some attribute is used and similarly another relevant detail is plotted on y-axis. The plus sign (+) denotes one of the two classes and the bubble sign (O) denotes another class. This is an example of a linearly separable problem.

The dataset for building and training such ML classifiers to solve the fault-prediction problem are also available on websites. AR1, AR2, AR3, AR4, AR5 and AR6 are examples of widely used dataset in building such classifiers. The most popular ML techniques for implementing the classification algorithms are decision tree (DT), artificial neural network (ANN), support vector machine (SVM) and convolution neural network (CNN).

4.2.1.2 Quantity Prediction as Regression Problem

The quantity aspect of software measurements is directly associated with the size of the software, which is in turn associated with the effort, cost and duration of the entire software project. Hence, the quantity-based software measurement is modeled as an effort estimation problem. Some relevant features of similar past projects are selected and fed to the ML-based model so that it can estimate the cost or efforts required to complete the current project. The desired output is over a range of numbers, hence, the problem can be formulated as a regression-based learning problem.

Let x_1 be the attribute set of the project on which the effort (to be estimated) depends. This attribute set forms the set of independent variables, and the effort (y) is the dependent variable. The regression-based estimation basically depends on the dataset of previous projects. This past dataset serves as a training dataset for our model. Then, some ML-based regression technique fits an equation to the training

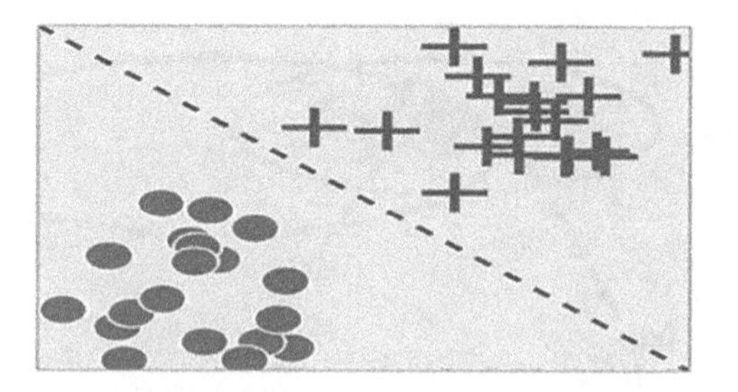

FIGURE 4.2 Classification [Class 0: not so risky; Class +: risky].

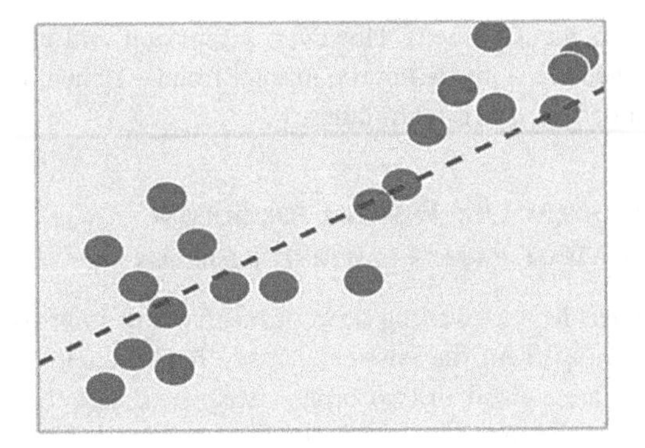

FIGURE 4.3 Regression.

data and yields y in terms of x_1. It can be formulated as Equation (4.1). The linear regression can be exemplified as in Figure 4.3 where the dotted line is the function fitted by some ML-based regression algorithm and bubbles denote the data-points from a past project dataset.

$$y = b_0 + b_1 * x_1 \tag{4.1}$$

Where

y is the dependent variable,
b_0 is the constant,
b_1 is the coefficient and
x_1 is the independent variable.

The dataset for building and training such ML-based estimation model to solve the effort estimation problem are also available on websites. COCOMO’81 and ISBSG are examples of widely used datasets in building such predictors. The most popular ML techniques for implementing the regression algorithms are DT, ANN, SVM and CNN.

In this section, we have successfully learned about how to formulate the software measurement as a learning problem. The quality-based measurement has been formulated as a classification problem, the fault-prediction problem. Similarly, the quantity-based measurement of the software has been formulated as a regression problem, effort estimation problem.

Now, the problem has been defined clearly. Next, we have to go through the available ML techniques in detail, so that we can select the most suitable one to solve our formulated problems.

Before moving to the next subsection for ML techniques, it is to be noted that both classification and regression are supervised leaning methods. In supervised methods, the predicted or estimated values are obtained by mapping the input–output pairs provided from the past labeled dataset. Another class of ML algorithms is for unsupervised type of problems where we only have input data-points (unlabeled) and we need to explore patterns hidden in that huge dataset; this is called clustering. The clustering algorithms, such as self-organizing map, are also finding applications in the field of software measurement. However, regression and classification (supervised ML) techniques are dominating the market trends. Hence, in this chapter, our focus will be on supervised ML techniques.

4.2.2 ML Methods at Our Disposal for Solving Software Measurements Learning Problems

Tom Mitchell defined ML as writing a program that can improve its performance P at task T by learning from the past experience E (Mitchell 1997). Overall, ML techniques can be categorized in two broad categories, namely, supervised learning techniques (regression and classification) and unsupervised learning techniques (clustering, dimension reduction and association) (Witten et al 2011; Ethem Alpaydin 2014), as shown in Figure 4.4.

Supervised learning algorithms make predictions for unseen data using the provided labeled training data (input–output pair). Unsupervised learning techniques find hidden patterns in the provided unlabeled training dataset. Our focus is on supervised learning techniques in this chapter. Supervised learning includes two major categories, namely, classification and regression techniques.

Classification is a ML approach which is supervised in nature. In classification, a computer program learns from the data fed to the program itself, and then using the learned knowledge, it classifies unforeseen new instances. Classification can be two-class classification, that is, identifying whether the module is defective or non-defective. Classification can be multiclass classification as well, such as classifying the module under three classes, namely, low, medium and high. Classification techniques find applications in speech recognition, handwriting recognition, biometric surveillance system and document classification, among others.

In empirical software measurement, fault prediction is formulated as a two-class classification problem. As shown in Figure 4.5, If $x1 > \theta 1$ and $x2 > \theta 2$, then low risk, otherwise high-risk. Low-risk class is denoted with a plus sign (+) and high-risk class is denoted with a minus sign (–).

Regression is an ML approach which is supervised in nature. In regression, a computer program learns from the data fed to the program itself, and then using the learned knowledge, it predicts output which is continuous for the unseen input. In empirical software measurement, effort estimation problem is formulated as a regression-based supervised learning problem. The effort is treated as a dependent

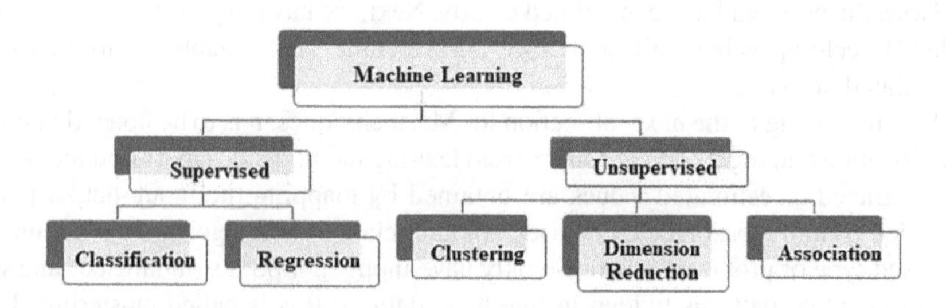

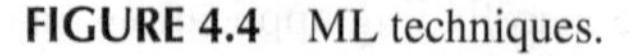
FIGURE 4.4 ML techniques.

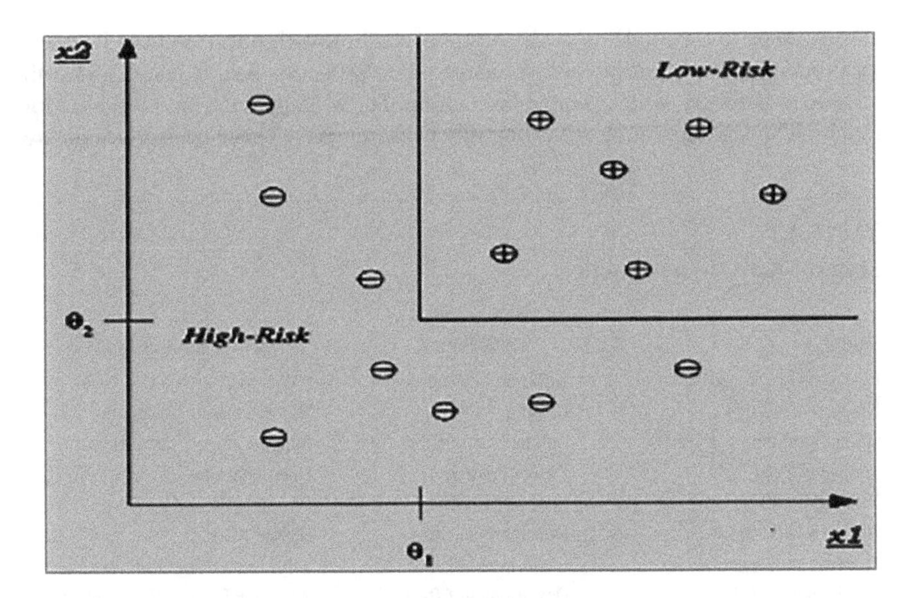

FIGURE 4.5 Fault prediction as a classification problem.

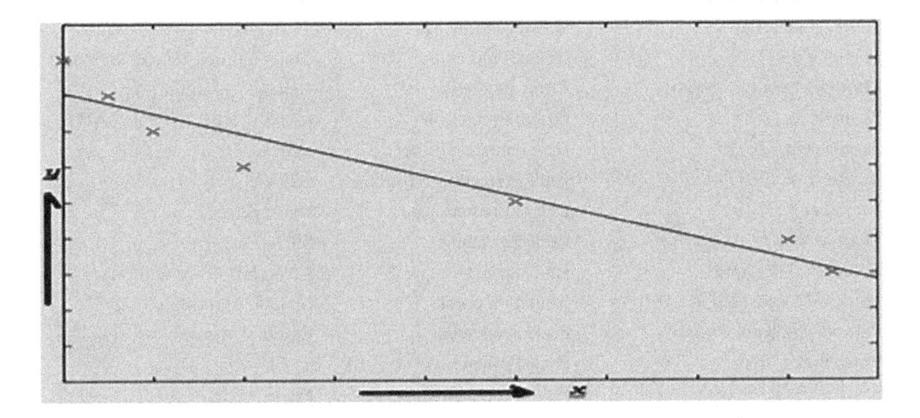

FIGURE 4.6 Effort estimation as regression.

variable and some selected attributes of the project are taken as independent variables. Training set {(*xn*, *yn*)}*Nn*=1 where the label for a pattern *xn* ∈ RD is a real value *yn* ∈ R, as shown in the Figure 4.6.

4.3 CURRENT TRENDS OF USING ML IN SOFTWARE MEASUREMENTS

This section presents a review of the current state-of-the-practice of ML in software measurements. ML is like the other half of the existing approaches to software measurements and completes in the terms of dealing with uncertainties. The current trend is summarized in Table 4.1. Since the 1980s, this field is resonating with software researchers, which is quite fascinating. The ML approach has been used for prediction and/or estimation regarding the measurements. These measurements are

TABLE 4.1
State-of-Art-Of-The-Practice

Study	Measurement Contributed	ML Technique Used
Pospieszny et al. (2018)	Effort and Duration	SVM, MLP–ANN and GLM
Hosni et al. (2017)	Effort	Heterogeneous Ensembles
Gárcía Floríano et al. (2018)	Enhancement Effort	Support Vector Regression
Tong et al. (2017)	Defect Prediction	Deep Learning
Yang et al. (2017)	Defect Prediction	Ensemble Learning
Araujo et al. (2017)	Development Effort	Hybrid MLPs
Moosavi et al. (2017)	Development Effort	Satin Bowerbird Optimizer
Jodpimai et al. (2018)	Development Effort	RA, CART, SVR, RBF and kNN
Nizamani et al. (2017)	Enhancement Requests	BN, SVM, Logistic Regression
Miholca et al. (2018)	Defect Prediction	GRARs–ANN
Jayanthi et al. (2018)	Defect Prediction	ANNs
Tomar et al. (2018)	Quality Prediction	TLBO–ANN
Li et al. (2017)	Defect Prediction	Convolutional Neural Network
Murillo-Morera et al. (2017)	Effort Prediction	Genetic Algorithm
Zhang et al. (2018)	Defect Prediction	LR, BN, RBF, MLP and ADT
Benala et al. (2017)	Development Effort	DABE
Kumar et al. (2017)	Fault Prediction	LSSVM
Arar et al. (2017)	Defect Prediction	Naïve Bayes
Moussa et al. (2017)	Fault-Proneness	PSO–GA
Boucher et al. (2018)	Fault-Proneness	ANN, NB, SVM and DT
Murillo-Morera et al. (2016)	Fault-Proneness	Genetic Algorithm
Rathore and Kumar (2016)	Fault-Proneness	Ensemble Learning
Rong et al. (2016)	Fault-Proneness	SVM
Ryu and Baik (2016)	Fault-Proneness	Naïve Bayes
Satapathy et al. (2016)	Effort Prediction	Tree-based Learning
Azzeh et al. (2017)	Effort Prediction	ANN
Chen et al. (2017)	SBSE	SBSE
Huang et al. (2017)	Missing Value	kNN
Rathore and Kumar (2017)	Fault Prediction	Ensemble Learning
Yu et al. (2017)	Defect Prediction	Feature Selection
Zhou et al. (2017)	Defect Prediction	Tree-based Learning
Manjula and Florence (2018)	Defect Prediction	Deep Learning
Tantithamthavorn et al. (2018)	Fault Prediction	Parameter Optimization
Vig and Kaur (2018)	Test Effort Estimation	MLP–ANN
Viji et al. (2018)	Fault Prediction	ANN–SOM
Turabieh et al. (2019)	Fault Prediction	Binary+(GA/PSO/ACO)

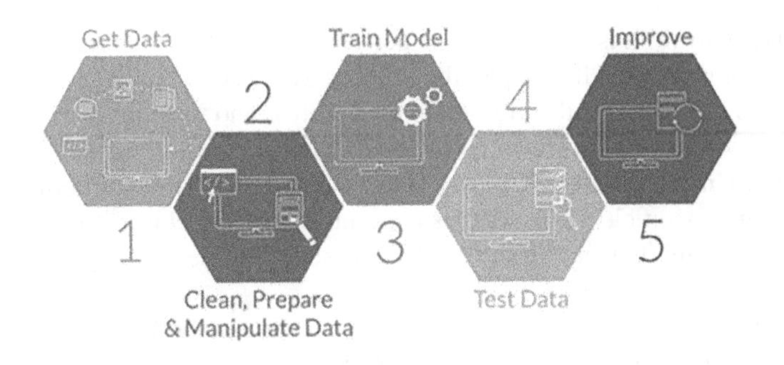

FIGURE 4.7 Steps to build an ML model for software measurements.

for both the internal and external attributes of software. This includes the measurement of software quality, size of the software, the development cost of the software, the effort incurred in the project or software, effort for the maintenance task, identification of the defective module, as well as the time or cost incurred in testing the software under development or after release.

The ML model is deployed in situations where we do not have the solution algorithm for problem-solving. In such situations, either we have little knowledge or no knowledge of the problem. In such cases, the data fills the gap between the problem and the solution. We collect the data, train our model and make our predictions and/or estimations using the trained model for unforeseen instances later. The process to build an ML-based model is depicted in Figure 4.7.

4.5 CRITERIA FOR MODEL EVALUATION

The evaluation of the performance of individual models is made over specific criteria. The criteria vary with the type of the problem. In the case of a classification-type problem, the output is discrete. In a regression-type problem, the output is continuous. Hence, we discuss the criteria for both cases in separate subsections below.

4.5.1 Software Quality Measurement Model (Classification)

The evaluation of the performance of predictors (for classifiers) is made using one or more of the following criterion (Hanley and McNeil 1982; Lehmann and Romano 2008; Ross 2005):

- **Confusion Matrix**: This is a two-dimensional matrix that represents the information about the correctly classified and wrongly classified data-points by the candidate prediction model.
- **Sensitivity**: It can be defined as the percentage of the "faulty" modules over the total number of modules which are correctly classified by the model.
- **Specificity**: It can be defined as the percentage of the "safe" modules over the total number of modules which are correctly classified by the proposed model.

- **ROC Curve**: It is to measure the performance of the prediction model and present it graphically. To plot the ROC curves, a sufficient number of cutoff points between 0 and 1 are randomly selected, and then, the sensitivity and specificity at each selected cutoff point is calculated and marked.
- **Area Under the ROC Curve (AUC)**: It is a combination of both sensitivity and specificity. The AUC is computed, and the closer it is to the unit area, the more accurate the model is assumed to be (Hanley and McNeil 1982).
- **Accuracy**: It measures how correct is the prediction of the proposed model. It is a ratio of the number of instances which are correctly classified to the total number of instances.

All the above criteria are popular in the literature for comparing the performance of prediction models and to determine the best model (Gondra 2008; Dejaeger et al. 2013; Malhotra 2014; Czibula et al. 2014).

4.5.2 Software Quantity Measurement Model (Regression)

The performance of effort estimation model is evaluated using magnitude of relative error (MRE) and mean magnitude of relative error (MMRE) metrics (Pospieszny et al. 2018) The MRE can be computed for a specific project instance using Equation 4.2.

$$\text{MRE} = |\,\text{Effort}_{(\text{actual})} - \text{Effort}_{(\text{predicted})}\,| \div \text{Effort}_{(\text{actual})} \tag{4.2}$$

Where the $\text{effort}_{(\text{actual})}$ is the effort provided in the dataset for 62 project instances, and $\text{effort}_{(\text{predicted})}$ is the effort predicted by the proposed model for that specific project instance.

The is the averaged value of MRE over the entire dataset and can be computed using Equation (4.3) as follows:

$$\text{MMRE} = \sum \text{MRE} \div \text{Num} \tag{4.3}$$

Where the MRE over all the projects are summed up and Num represents the project count in the dataset.

Different criteria are used in different scenarios. In the next section, we will discuss a case study covering the major aspects of software measurements. The performance evaluation criteria are used to measure the performance of the individual model.

4.6 CASE STUDY

This presents an actual implementation of ML techniques for software measurements. The case study deploys neural network for prediction and estimation of software measurements.

4.6.1 ML Technique for Measuring Software Quality – A Case Study

Measuring software quality is a very difficult but essential task for the successful delivery of the product on time. In this case study, we consider quality prediction as a classification task. For the sake simplicity, we have considered it as a two-class classification problem, in which the software modules are classified as "Fault-prone" or "Safe" modules. The fault-prone modules are risky and prone to errors. The safe modules are not so risky in the final product. The aim is to predict the faulty modules in the early development phases so that testing efforts can be targeted to those faulty modules and the quality of the product can be improved.

Six models are developed using ANN over six datasets (CM1, KC1, KC2, PC1, JM1 and combined one). The performance of all the proposed models is evaluated using ROC, AUC and accuracy as performance evaluation criteria. The complete experimental flow is modeled in Figure 4.8. First, the data is collected, and then the features are selected. Next, the models are trained and tested. Finally, the performance is measured for all six models over the six selected datasets.

4.6.1.1 Data Description

The case study has been carried out using NASA datasets (CM1, KC1, KC2, PC1, JM1 and a combination of CM1, KC1, KC2, PC1, JM1) available in the PROMISE repository (Sayyad and Menzies, 2005; PROMISE; Thomas 1976; Goyal and Parashar 2018). Figure 4.9 shows the class distribution over all the selected datasets.

All six models are developed using standard implementation of MATLAB software using 10-fold cross-validation. Confusion matrix is used to demonstrate the prediction strength of classifiers. Figure 4.10 shows the prediction power of model for KC1 dataset.

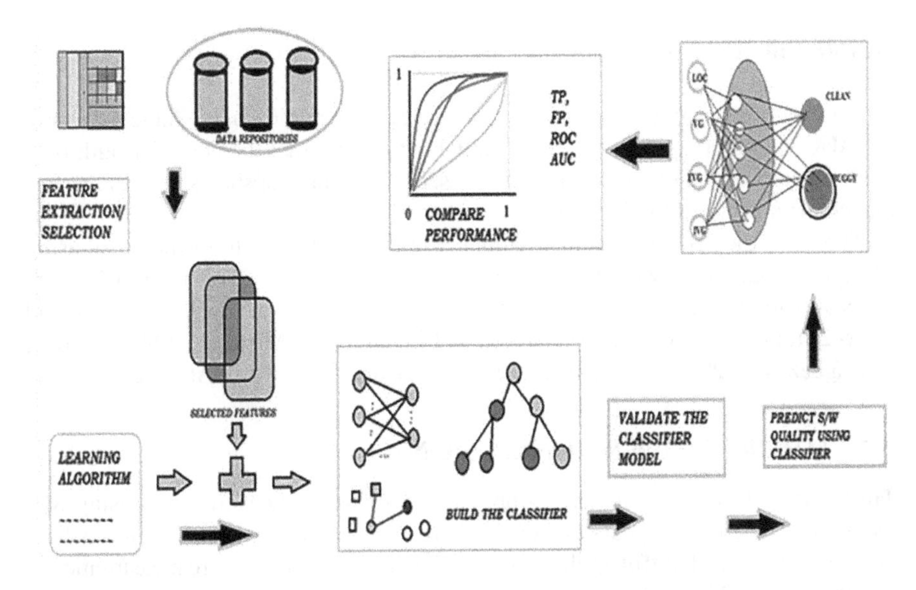

FIGURE 4.8 Experimental set-up.

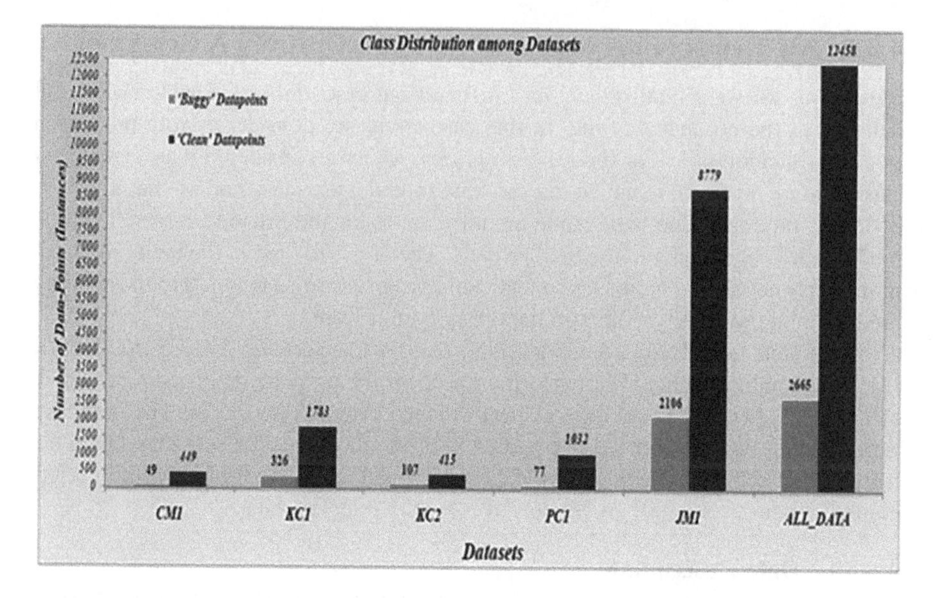

FIGURE 4.9 Datasets—class distribution as "Buggy"–"Clean" classes.

OUTPUT	ANN	
BUGGY	43	283
CLEAN	37	1746
	BUGGY	CLEAN
	TARGET CLASS	

FIGURE 4.10 Confusion matrix—KC1 dataset.

Another criteria for performance measurement is the AUC curve. Table 4.2 shows the AUC values computed for all six models. It is clear that the classifier built over the KC2 dataset and the neural network-based ML technique shows the best performance, with a value of 0.8315 for AUC.

The case study is designed in alignment with the current state-of-the-art of using ML for software quality prediction. A broad range of literature from the last 30 years advocates for the use of ML for software empirical measurements (Boucher and Badri 2018; Gyimothy et al. 2005; Zhou and Leung 2006; Mishra and Shukla 2012; Rodríguez et al. 2013; Laradji et al. 2015; Erturk and Sezer 2015; Abaei 2015).

4.7 FUTURE SCOPE & CONCLUSION

This chapter brings an empirical approach to software measurements using ML techniques. The past 30 years of work done by thousands of researchers in this field are evidence of the fruitful application of ML in the field of software measurements. We introduced ML techniques and their applications in software measurements

TABLE 4.2
AUC Measure

AUC	ANN
CM1	0.7286
KC1	0.7878
KC2	0.8315
PC1	0.7187
JM1	0.7102
ALL_DATA	0.7282

empirically. Then, the research carried out in the field of empirical software measurements using ML techniques from 1988 through 2019 was discussed. The two most important software attributes to be measured are software quality and software quantity. From the research trends, it was observed that the most popular technique is neural network, which has been used frequently for both quality and quantity measurements considering effort estimation and fault prediction as exemplary problems. Hence, model development was explained using neural networks as an example. Then, the measurement criteria were explained for both regression and classification problems. Case studies (Goyal and Bhatia 2019a, 2019b) were also included as an important part of this chapter to provide practical orientation to readers about the implementation of the ML technique to software measurements. Finally, we concluded the work with special references to deep learning (Goodfellow et al. 2016) as a tool to perform software measurements in future with LSTM or CNN architecture.

From the systematic study and the review conducted, it can be concluded that for the past three decades, very crucial research has been performed in the field of software measurement using ML algorithms. The sole aim is to handle the discrepancies in old-fashioned estimation and measurement techniques, which can improve the project completion success rate. It also aims to incorporate the modern software development and management approaches along with the existing approaches.

ML techniques are found to be the most suitable for the current software development environment. To accurately measure the software attributes (processes, products or resources), ML techniques are used, resulting in increased quality of the product. Further, improvements with attribute selection and/or feature extractions and optimization of parameters of the predictors required need to be considered effectively.

REFERENCES

Abaei, G., A. Selamat and H. Fujita. 2015. An empirical study based on semi-supervised hybrid self-organizing map for software fault prediction. *Knowl. Based Syst.* 74: 28–39.

Arar, O.F. and K. Ayan. 2017. A feature dependent Naive Bayes approach and its application to the software defect prediction problem. *Appl. Soft Comput.* 59: 197–209.

Araujo, A., L.I. Oliveira and S. Meira. 2017. A class of hybrid multilayer perceptrons for software development effort estimation problems. *J. Expert Sys. App.* 90C: 1–12. doi: 10.1016/j.eswa.2017.07.050.

Azzeh, M., A.B. Nassif and S. Banitaa. 2018. Comparative analysis of soft computing techniques for predicting software effort based use case points. *IET Softw.* 12.1: 19–29.

Benala, T.R. and R. Mall. 2018. DABE: Differential evolution in analogy-based software development effort estimation. *Swarm Evol. Comput.* 38: 158–172.

Boucher, A. and M. Badri. 2018. Software metrics thresholds calculation techniques to predict fault-proneness: An empirical comparison. *Inf. Softw. Technol.* 96: 38–67.

Chaos Report, 2015. The Standish Group.

Chen, J., V. Nair and T. Menzies. 2017. Beyond evolutionary algorithms for search-based software engineering. *Inf. Soft. Technol.* doi: 10.1016/j.infsof.2017.08.007.

Czibula, G., Z. Marian and I. G. Czibula. 2014. Software defect prediction using relational association rule mining. *Inf. Sci.* 264: 260–278.

Demarco, T. 1982. *Controlling Software Projects: Management, Measurement and Estimation.* Yourdon Press.

Dejaeger, K., T. Verbraken and B. Baesen. 2013. Toward comprehensible software fault prediction models using Bayesian network classifiers. *IEEE Trans. Soft. Eng.* 39.2: 237–257.

Erturk, E. and E. A. Sezer. 2015. A comparison of some soft computing methods for software fault prediction. *Expert Syst. Appl.* 42: 1872–1879.

Ethem Alpaydin. 2014. *Introduction to Machine Learning (Adaptive Computation and Machine Learning Series.* MIT Press.

García-Floriano, A., C. López-Martín, C. Yáñez-Márquez and A. Abran. 2018. Support vector regression for predicting software enhancement effort. *Inf. Softw. Technol.* 97: 99–109.

Gondra, I. 2008. Applying machine learning to software fault-proneness prediction. *J. Syst. Softw.* 81.5: 186–195.

Goodfellow, I., Y. Bengio and A. Courville. 2016. *Deep Learning.* MIT Press.

Goyal, S. and A. Parashar. 2018. Machine learning application to improve COCOMO model using neural networks. *Int. J. Info. Technol Comp. Sci. IJITCS).*10 3: 35–51. doi: 10.5815/ijitcs.2018.03.05.

Goyal, S. and P. K. Bhatia. 2019a. A non-linear technique for effective software effort estimation using multi-layer perceptrons. *International Conference on Machine Learning, Big Data, Cloud and Parallel Computing* (ComITCon 2019). IEEE. 978-1-7281-0212-2.

Goyal, S. and P. K. Bhatia. 2019b. Feature selection technique for effective software effort estimation using multi-layer perceptrons. *Proceedings of ICETIT 2019, Emerging Trends in Information Technology.* doi: 10.1007/978-3-030–30577-2_15

Gyimothy, T., R. Ferenc and I. Siket. 2005. Empirical validation of object-oriented metrics on open source software for fault prediction. *IEEE Trans. Software Eng.* 31: 897–910.

Hanley, J. and B. J. McNeil. 1982. The meaning and use of the area under a Receiver Operating Characteristic ROC curve. *Radiology* 143: 29–36.

Hosni, M., A. Idri and A. Abran. 2017. Investigating heterogeneous ensembles with filter feature selection for software effort estimation. *Proceedings of the 27th International Workshop on Software Measurement and 12th International Conference on Software Process and Product Measurement on- IWSM Mensura'17:* 207–220.

Huang, J., J. W. Keung, F. Sarro et al. 2017. Cross-validation based K nearest neighbor imputation for software quality datasets: An empirical study. *J Syst. Soft.* doi: 10.1016/j.jss.2017.07.012.

Jayanthi, R. and L. Florence. 2018. Software defect prediction techniques using metrics based on neural network classifier. *J. Cluster Comput.* Special Issue 1: 1–12.

Jodpimai, P., P. Sophatsathit and C. Lursinsap. 2018. Re-estimating software effort using prior phase efforts and data mining techniques. *Innov. Syst. Soft. Eng.* 14 3: 209–228.

Kumar, L., S. K. Sripada and S.K. Rath. 2018. Effective fault prediction model developed using Least Square Support Vector Machine (LSSVM). *J. Syst. Soft.* 137: 686–712.

Laradji, I. H., M. Alshayeb and L. Ghouti. 2015. Software defect prediction using ensemble learning on selected features. *Inf. Soft. Technol.* 58: 388–402.

Lehmann, E. L. and J. P. Romano. 2008. *Testing Statistical Hypothesis: Springer Texts in Statistics.* Springer, New York, 3rd ed. 2005. Corr. 2nd printing 2008 edition (10 September 2008) ISBN-13: 978–0387988641.

Li, J., P. He, J. Zhu, and M. Lyu. 2017. Software defect prediction via convolutional neural network. *IEEE International Conference on Software Quality, Reliability and Security,* 318–328.

Malhotra, R. 2014. Comparative analysis of statistical and machine learning methods for predicting faulty odules, *Appl. Soft Comp.* 21: 286–297.

Manjula, C. and L. Florence. 2018. Deep neural network based hybrid approach for software defect prediction using software metrics. *Cluster Comp.* doi: 10.1007/s10586-018-1696-z.

Miholca, D., G. Czibula and I. G. Czibula. 2018. A novel approach for software defect prediction through hybridizing gradual relational association rules with artificial neural networks. *Info. Sci—Info. Comp. Sci. Intel. Syst., App. Int. J.* 441 C: 152–170.

Mishra, B. and K. Shukla. 2012. Defect prediction for object oriented software using support vector based fuzzy classification model. *Int. J. Comput. Appl.* 60 15: 8–16.

Mitchell, T. 1997. *Machine Learning.* McGraw-Hill.

Moosavi, S. and V. Bardsiri. 2017. Satin bowerbird optimizer: A new optimization algorithm to optimize ANFIS for software development effort estimation. *Eng. App. Arti. Intel.* 60: 1–15.

Moussa, R. and D. Azar. 2017. A PSO-GA approach targeting fault-prone software modules. *J. Syst. Soft.* 132: 41–49.

Murillo-Morera, J., C. Castro-Herrera and J. Arroyo. 2016. An automated defect prediction framework using genetic algorithms: A validation of empirical studies. *Inteligencia Artificial.* 19 57: 114–137. doi: 10.4114/ia.v18i56.1159.

Murillo-Morera, J., C. Quesada-López and C. Castro-Herrera. 2017. A genetic algorithm based framework for software effort prediction. *J. Softw. Eng. Res. Dev.* 5: 1–4.

Nizamani, Z. A., H. Liu and Z. Niu. 2018. Automatic approval prediction for software enhancement requests. *J. Autom. Softw. Eng.* 25 2: 347–381.

PROMISE. http://promise.site.uottawa.ca/SERepository.

Pospieszny P., B. Czarnacka-Chrobot and A. Kobylinski. 2018. An effective approach for software project effort and duration estimation with machine learning algorithms. *J. Syst. Soft.* 137: 184–196.

Rathore, S. S., and S. Kumar. 2017. Linear and non-linear heterogeneous ensemble methods to predict the number of faults in software systems. *Know.-Based Syst.* 119: 232–256.

Rathore, S. S. and S. Kumar. 2017. Towards an ensemble based system for predicting the number of software faults. *Expert Syst. App.* 82: 357–382.

Rodríguez, D., R. Ruiz and J. C. Riquelme. 2013. A study of subgroup discovery approaches for defect prediction. *Info, Soft. Technol.* 55 10: 1810–1822.

Rong, X., F. Li and Z. Cui. 2016. A model for software defect prediction using support vector machine based on CBA. *Int. J. Intel. Syst. Technol App.* 15 1: 19–34.

Ross, S.M. 2005. *Probability and Statistics for Engineers and Scientists.* 3rd Ed., Elsevier Press. ISBN:81-8147-730-8.

Ryu, D. and J. Baik. 2016. Effective multi-objective naïve Bayes learning for cross-project defect prediction. *J. App. Soft Comp.* 49 C: 1062–1077.

Satapathy, S., B. Achary and S. Rath. 2016. Early stage software effort estimation using random forest technique based on use case points. *IET Soft. Inst. Eng. Technol.* 10 1: 10–17.

Sayyad, S. and T. Menzies. 2005. *The PROMISE Repository of Software Engineering Databases.* University of Ottawa, http://promise.site.uottawa.ca/SERepository.

Tantithamthavorn, C., S. Mcintosh and A. Hassan. 2017. An empirical comparison of model validation techniques for defect prediction models. *IEEE Trans. Soft. Eng.* 43 1: 1–18.

Thomas, J. 1976. McCabe, a complexity measure. *IEEE Trans. Soft. Eng.* 2 4: 308–320.

Tomar, P, R. Mishra and K. Sheoran. 2018. Prediction of quality using ANN based on teaching-learning optimization in component-based software systems. *J. Soft. Pract. Exper.* 48 4: 896–910.

Tong, H., B. Liu and S. Wang. 2018. Software defect prediction using stacked denoising autoencoders and two-stage ensemble learning. *Inf. Soft. Technol.* 96: 94–111.

Turabieh, H., M. Mafarja and X. Li. 2019. Iterated feature selection algorithms with layered recurrent neural network for software fault prediction. *Exper. Syst. App.* 122: 27–42. doi: 10.1016/j.eswa.2018.12.033.

Vig, V. and A. Kaur. 2018. Test effort estimation and prediction of traditional and rapid release models using machine learning algorithms. *J. Intel. Fuzzy Syst. IOS Press.* 35: 1657–1669. doi: 10.3233/JIFS–169703.

Viji, C., N. Rajkumar and S. Duraisamy. 2018. Prediction of software fault-prone classes using an unsupervised hybrid SOM algorithm. *Cluster Comp.* doi: 10.1007/s10586-018-1923-7.

Witten, L. H., E. Frank and M. A. Hell. 2011. Data mining: Practical machine learning tools and techniques In *Acm Sigsoft Software Engineering Notes*, 3rd ed. Burlington: Morgan Kaufmann, 90–99.

Yang, X., D. Lo, X. Xin and J. Sun. 2017. TLEL: A two-layer ensemble learning approach for just-in-time defect prediction. *J. Info. Soft. Technol.* 87: 206–220.

Yu, Q., S. Jiang and Y. Zhang. 2017. A feature matching and transfer approach for cross-company defect prediction. *J. Syst. Soft.* 132: 366–378 doi: 10.1016/j.jss.2017.06.070.

Zhang, Y., D. Lo, X. Xia, and J. Sun. 2018. Combined classifier for cross-project defect prediction: An extended empirical study. *Front. Comput. Sci.* 12 2: 280–296.

Zhou, L., R. Li, S. Zhang and H. Hang. 2017. Imbalanced data processing model for software defect prediction. *Wireless Pers Commun.* 6: 1–14. doi: 10.1007/s11277-017-5117-z.

Zhou, Y. and H. Leung. 2006. Empirical analysis of object-oriented design metrics for predicting high and low severity faults. *IEEE Trans. Soft. Eng.* 32: 771–789.

5 Project Estimation and Scheduling Using Computational Intelligence

Vikram Bali
JSS Academy of Technical Education

Shivani Bali
Lal Bahadur Shastri Institute of Management

Gaurav Singhania
JSS Academy of Technical Education

CONTENTS

5.1 INTRODUCTION

Humans respond to a situation based on the intelligence they develop over time, either by understanding from situations encountered previously or by facts provided for an existing situation. Therefore, human intelligence is basically an action generated by the human being for the situation created. For example, if a person is taught addition, the mind is trained for the process of addition; and if the same process is encountered again, the person will apply that process to generate the result. We call this process intelligence. This intelligence can be generated from many new aspects of a process, for example, if you repeat a similar process, you will take a shorter time for its completion. In the second attempt, you will be aware of some of the risks and will know how to tackle them.

However, the main questions that arise are the following:

1. How will you make similar intelligence for machines, or if this intelligence can be imparted to machines or not?
2. What do you mean by intelligence?
3. Is intelligence defined by a basic understanding of the situation or by just following the facts?
4. How will you define computational intelligence in terms of human behavior?
5. How is this related to other fields of computer science?
6. Why is computational intelligence developed even though artificial intelligence has been developed?

Although computational intelligence has many paradigms, here we will discuss artificial neural networks (ANNs), fuzzy system (FS), evolutionary computation (EC) and swarm intelligence (SI).

This chapter aims to propose a basic model to generate better scheduling techniques and how it is used in computational intelligence.

5.2 PROPOSED MODEL

In the proposed model, we are trying to implement the project scheduling technique, that is, a Gantt chart using the evolutionary algorithm. Here, we will fix some parameters of the Gantt chart, and then we will present those parameters as the basis of the algorithm. Subsequently, we will try to establish that the results which are not being generated are more optimized than the previously generated results without using the evolutionary technique.

Consider an example of developing a software which has many tasks that need to be completed.

These tasks are defined, as shown in Figure 5.1:

- Market research
- Define specifications
- Overall architecture
- Project planning

SOFTWARE DEVELOPMENT PROJECT

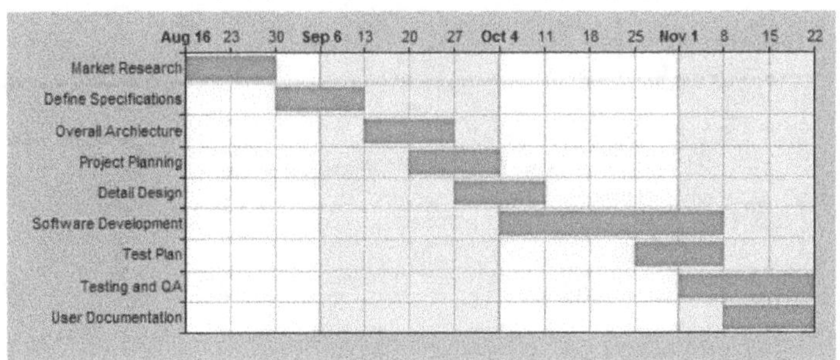

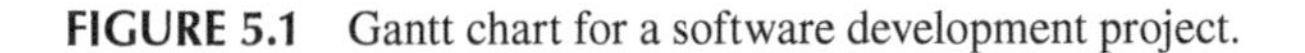

FIGURE 5.1 Gantt chart for a software development project.

- Detail design
- Software development
- Test plan
- Testing and quality analysis
- Documentation

We have created a Gantt chart for the software development project using a simple Gantt chart creator from Advanced Software Engineering Limited.

In this model, we will keep the Gantt chart parameters or the tasks fixed. Therefore, by applying the evolutionary algorithm on those parameters, this algorithm will generate optimized results.

In the above-created Gantt chart, we can see that there are multiple factors that depict the timeline of the Gantt chart. We will consider one of the factors and process it into the evolutionary model to get the optimized result.

We have considered the project planning schedule time which will undergo the evolutionary process. By considering one of the parameters of the Gantt chart, such as the project planning scheduler task time, which will undergo evolutionary algorithm, the main aim of this algorithm would be to reduce the expected time of this task to be completed, so that the overall time of the project is reduced. This Gantt chart was created using Ms Excel 2010.

5.2.1 Artificial Neural Networks (ANNs)

The term artificial intelligence has been coined from human intelligence. The human brain has a set of neurons for signal processing (Zhu 2014). The basic components of biological neurons (as shown in Figure 5.2) are dendrites, cell body, nucleus, axon and axon terminals. The dendrites help receive the electric signal from other neurons which flow from the cell body to the axon terminals which transmit the signal to the other neurons (Neves et al. 2018).

The model for the human neurons is the basis of artificial neurons (AN), which transmits the signal (Krasnopolsky 2013). Each AN receives a signal from the other neuron of the environment, gathers the signal, activates it and then transmits the signal to all connected neurons. Figure 5.3 represent a single neuron which comprises three

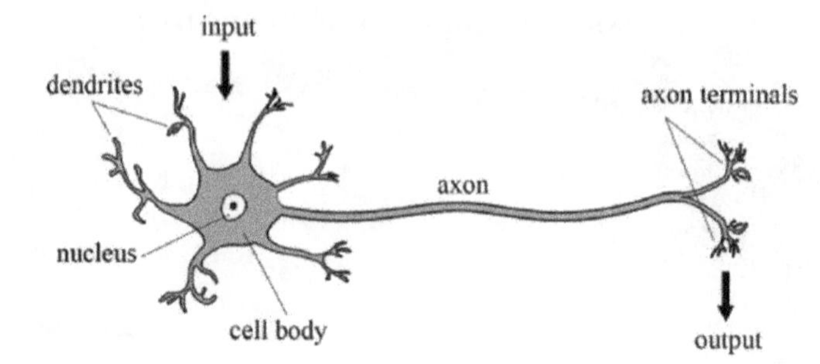

FIGURE 5.2 Biological neuron.

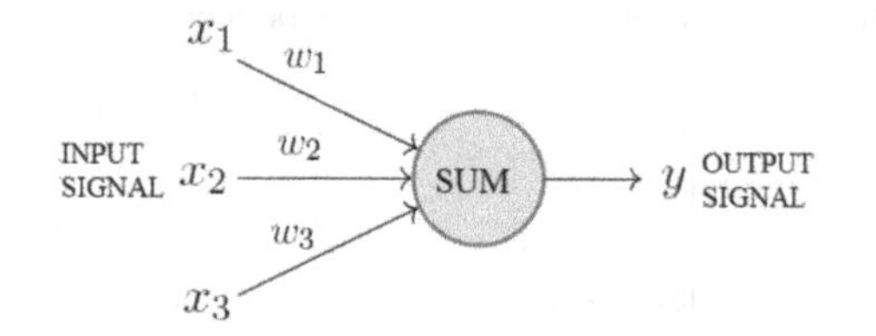

FIGURE 5.3 Artificial neuron.

components: input signal (x_1,x_2,x_3) having weights (w_1,w_2,w_3), computing function where the data is stored, and waiting for the activation function (Hagan et al. 1996).

The ANN is a network of multiple ANs. In ANNs, there are three layers: input layer, hidden layer and output layer (Duch 2003). A typical ANN is depicted in Figure 5.4.

Different ANNs are developed by simple combinations of artificial neurons, including:

- Hopfield networks or single-layered ANNs
- Standard backpropagation, product unit propagation or multilayered feed-forward ANNs
- Temporal ANNs like time-delayed neural networks
- Self-organizing ANNs like learning vector quantizer
- Combined supervised and unsupervised ANNs

They are used for a wide range of functions in the field of data mining, composing music, image processing, robot control, speech recognition, planning games, pattern recognition and many others (Huo et al. 2013).

5.2.2 Fuzzy System (FS)

The trivial fuzzy set system has only two values for decision making, which is either 0 or 1. As the entire situation can't be judged on the basis of these two values, we need a better method for judging situations based on other values which may interfere with the outcome of the situation.

In human intelligence, we can clearly see that it is being governed by other factors as well. For example, if you want to eat while doing some urgent work, you can

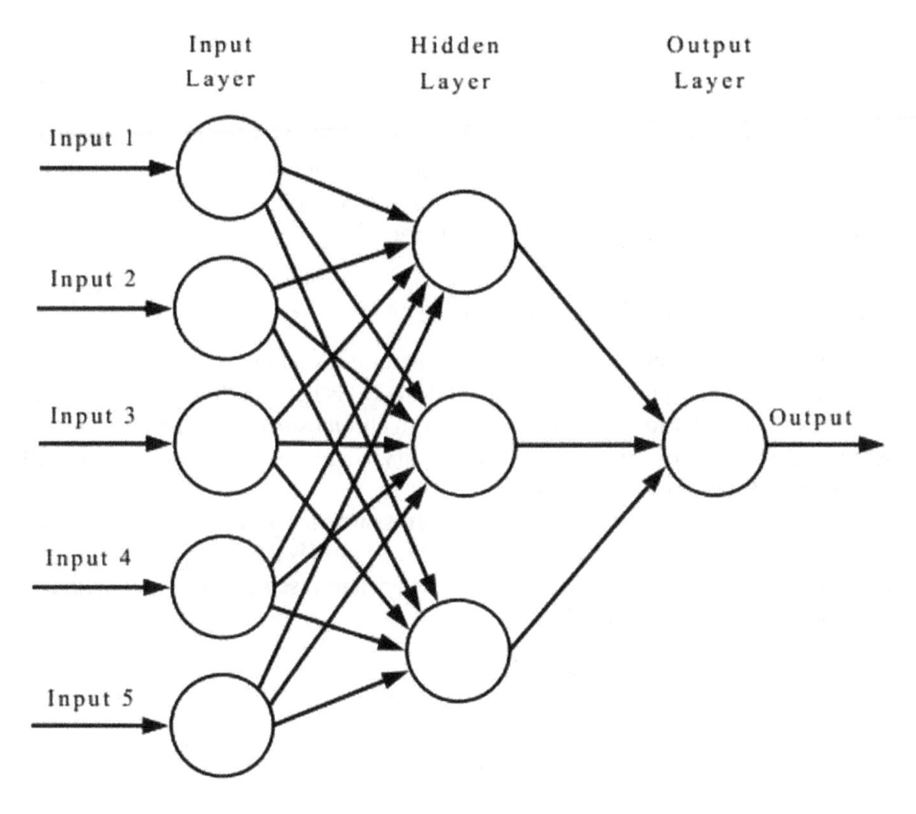

FIGURE 5.4 Artificial neural networks.

either deprioritize it so you can continue doing that work, or if you are very hungry, how will you decide what to do in a limited time frame? This is where the role of human intelligence comes into play. This kind of intelligence where there is no exact decision-making but a percentage is defined depending on foreign factors, the FS provides decisions which may be the most suitable according to the decisions based on human intelligence (Madani 2011).

The main aspects of using the FS are defined as below:

- Trivial decision-making is done only in binary, but the FS provides a further view of the decision depending on other factors interfering with the situation (Hagan et al. 1996).
- The FS helps to calculate the approximate values giving clarity in decision-making.
- FS allows logical reasoning keeping the uncertainty associated with the external factors (Zhu 2014).
- The decision-making in FS is quite accurate but the decisions taken by humans are not exact (Azar 2010).
- Example: Machine decision-making quality: binary (0/1).
- Human decision-making quality: present situation.

- FS decision-making quality: present situation+history and other factors depicted in percentage.

When fuzzy logic is introduced in the machine, the machine will be able to make good decisions. Today, fuzzy logic can be seen making decisions in many fields, for example, weather forecasting, traffic signaling, control system, home appliance, braking system, lift controlling, and many more.

5.2.3 Evolutionary Computation (EC)

The aim of EC is to imitate processes from natural evolution. As human evolution is based on the concept of survival of the fittest, EC is done in the same manner. In this process, we refer to each characteristic as a chromosome of individuals. As the set of individuals or population mutates, the surviving individual or population will have some different characteristics.

These remaining sets of individuals are crossover for a new set of individuals who are advanced from the previous set. As this process is done, a new generation is created, and this is repeated for more and more advanced sets of mutated individuals generated for survival (Konar 2005).

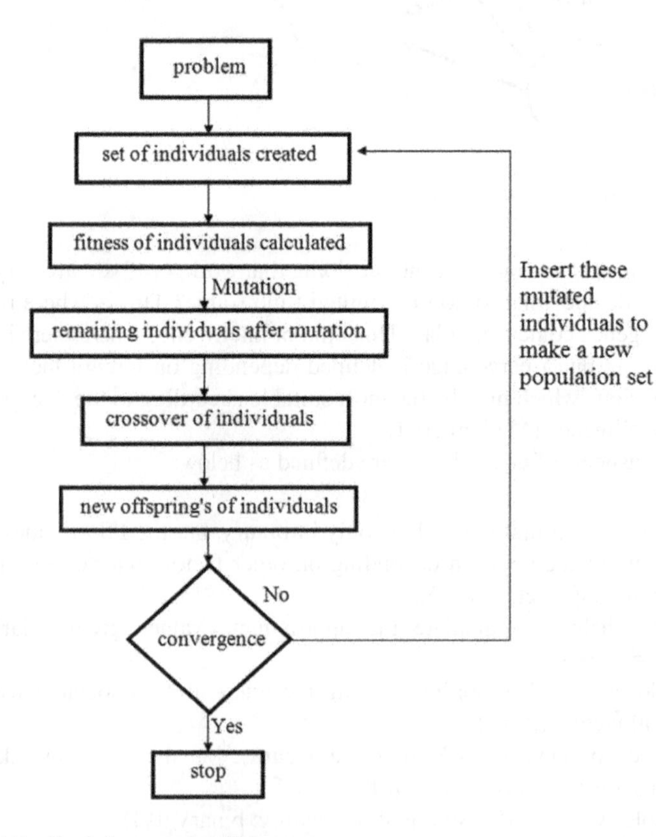

FIGURE 5.5 Evolutionary algorithm process.

In EC, this is done so that we come up with a solution of a set of problems with different situations.

Figure 5.5 depicts the general procedure for the evolutionary algorithm. The first step is the defined problem. The problem tells us about the characteristics or genes which are required for solving the problem. As the problem is defined, we generate a set of individuals or populations for processing.

All the chromosomes or characteristics of the individuals are calculated using a fitness function. The chromosomes are selected which are suitable for survival in the next generation. The selected chromosomes then undergo crossover to create a set of offspring. The newly created offspring have mixed characteristics who are inserted in the population.

As this is performed multiple times, that is, many generations of the children are passed, we can expect a composed solution for the set of individuals who survive, or the most optimal solution of the problem is generated. This is depicted in Figure 5.6.

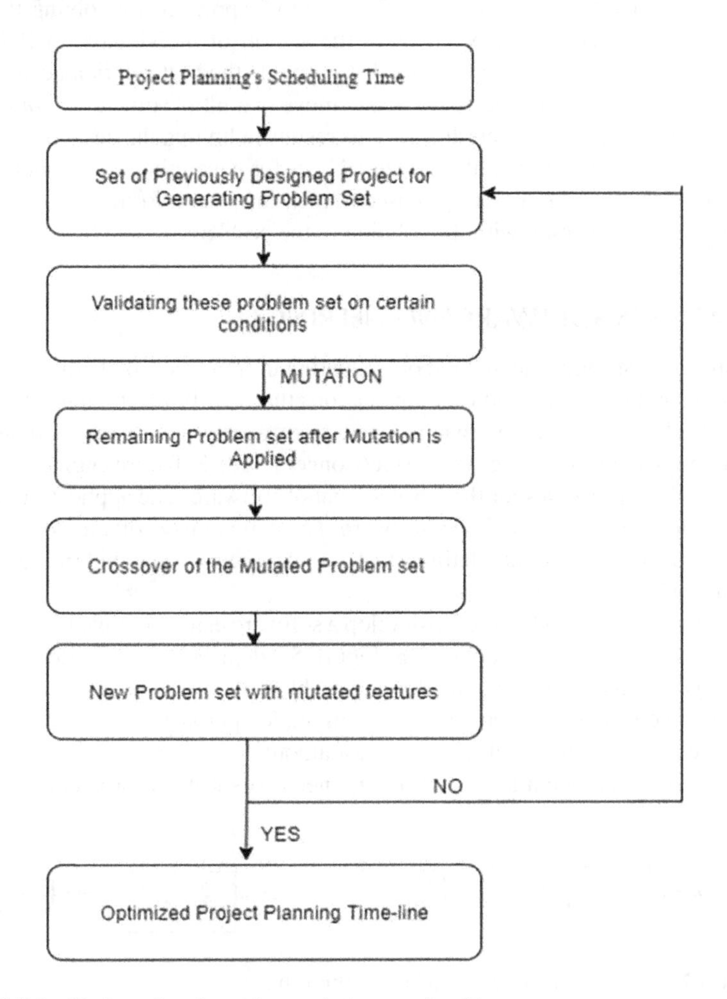

FIGURE 5.6 Project planning using evolutionary algorithm.

5.2.4 Swarm Intelligence (SI)

SI is developed from the local intelligence of organisms living or working in swarms. We can see how a swarm of fishes, bees, birds functions, and an example of collective intelligence is produced. SI refers to the intelligence developed by the brain of brains (Duch 2003). For example, if bees have to look for a new home, how they will find and choose the best one for them. In this case, more than 80% of the time bees make good decisions. This is done simply by signaling to other bees their decision by vibrating, or what we call a bee dance.

In reality, SI is very helpful which uses multiple processors for computing problems. Here, multiple brains develop an artificial swarm of brains, which works as a single unit and generates the most accurate decisions compared to the individual brain. Many studies have shown that SI plays a vital role in problem-solving (Liou 1990).

Particle swarm optimization (PSO) is a random approach for solving the problems, which is based on the behavior of the swarm of individuals. In PSO, each individual is called a particle which tends to provide the best solution according for the set of problems. The particles are free to move in multiple directions for the solution, adjusting their values accordingly. This results in finding the best position of the particle in the swarm compared to the neighbors for generating an optimal solution for the problem. A fitness function is used to measure the performance or the output of each particle in relation with the solution to the problem.

5.3 WHAT IS SOFTWARE ENGINEERING?

Software engineering is the method of evaluating user needs and planning, followed by creating and reviewing end user applications that will fulfill these needs through the use of software programming languages. A software is a set of instructions which are composed to generate a desired result (Conger 2008). Software engineering is the collection of the processes for the development of software. The application which is developed for the system is called a software. The main aim of software development is to make the process of completing a task simple, accurate and within the required time frame (Huges 1999).

Software engineering helps us to develop a software using multiple test conditions and validating them for its application. Figure 5.7 depicts the basic life cycle of an application which can be generated for any problem.

When we encounter a problem, we use multiple approaches to solve a problem, but we need to solve the problem in the least amount of time and with the minimum cost and maximum accuracy. For this, we need to solve the problem keeping in

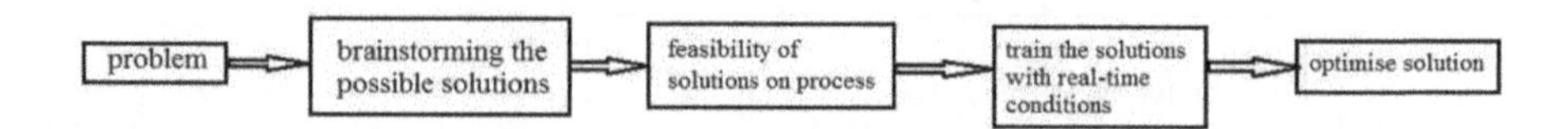

FIGURE 5.7 Software application cycle for a problem.

mind all the possible aspects which might affect the solution-generating process. Therefore, we need to check the solutions with the real-time possible conditions which might arise. This process continues, and after having multiple sets, we are left with the most optimized solution for the problem.

5.3.1 Software Engineering Problems

For any software project, the main aim is to develop that project in the allotted time frame. Some of the issues that the software scheduler faces while scheduling any software are the following:

- risk management of the task that might take place while the task is in completion.
- maintaining task integrity even if some external factors are affecting the task.
- while the scheduler is planning for the project, the scheduler should know about existing similar projects so that planning is done in a more efficient manner.
- lack of future aspects of scheduling is one of the main problems of software engineering project management.

5.3.2 Software Project Scheduling

Software project scheduling or project scheduling can be understood as the stepwise timely processing of the different modules in the development of a software keeping in mind the resources provided by the client and the developer's requirements (Conger 2008). The main aim of a project scheduler is to plan the software development process according to the resources provided (Pressman 2005). Scheduling can be classified into multiple phases.

1. **Select Project**: this is the initial stage for any project to be scheduled. In this phase, we have the basic knowledge of what we need.
2. **Aim of Project and Identify Its Scope**: when a project is developed, we look into the impact of the project during development and after the development. Once a project is selected, all the parties included equally agree to it, and all the possible results needed from the project are fulfilled.
3. **Project Infrastructure**: while developing the project, the project leader must have knowledge of an existing infrastructure into which the developing project can fit and must take a survey of the software and hardware standards which are required for the project.
4. **Project Characteristics Analysis**: the aim is to keep a watch on the methods and services used for development. Analysis is also done on the basis of implementing character, for example, if it is a product-driven or objective-driven development (Leung 2002).

5. **Project Product Activities and Product**: the product which will be used in the development of the project. The products that are to be used in the development are needed to be briefed on all aspects. The purpose of the product, relevant standards required for the project and quality criterion of the product used need to be clarified before they are used in the project.
6. **Effort Estimation of Activity**: all the activities which are going to be performed during the project need to be measured on the basis of the effort they will provide in the development cycle. Each and every activity has a role in either developing a phase or for validating the product being used in the development of the project (Pandian 2004).
7. **Identifying Activity Risk**: risk identification plays a vital role in the development of a project on many parameters. The activity-based risks are required to be planned in such a manner that they do not affect the overall process, or the degree of risk should be defined so that high-risk activities are implemented in such a manner that their impact on the overall development is minimized.
8. **Resource Allocation**: the resources provided for the development need to be allocated wisely. The capacity of the resources is first measured and then they are allocated so that we can have the best combination of all resources (Humphrey 1995). When the allocation is done, it needs to be done on revision basis to know whether the resources being applied give the required output. If some staff is needed for more than one task at the same time, then the resources need to be allocated on priority basis so that minimal processes wait for the resource.
9. **Review Plan**: when a scheduling task is being performed, we cannot keep a check on the completion of each and every process, so there is a risk that when the switching from one process to other process is done, it is necessary to keep a watch on the previously completed process, as well as to ensure that the output generated is sufficient for the next process to work upon.

5.3.3 Software Engineering Project Scheduling Technique

Gantt chart is a widely used technique used for the scheduling of the tasks while completing a project (Conger 2008). This is one of the simplest and effective techniques which can be used to schedule the tasks according to their requirements in project completion and using the resource in the most effective manner.

The Gantt chart comprises the following parts which are defined individually:

- Task list
- Task duration
- Resources
- Dependencies
- Budget
- Project completion date
- Milestones

Task list is one of the most important aspects of the Gantt chart as it contains all the important tasks of the scheduler which are required for the completion of the project.

Task duration is defined as the complete duration required by the task for completion, including the delays (Mukhopadhyay et al. 1992). There are two methods for calculating task duration:

I. **Analogous Estimating**: the task is compared to a previously completed task. For example, staff A will build the house in 20 days.
II. **Parametric Estimating**: the task is compared to a unit rate completion in unit time. For example, staff A will build walls at a rate of five walls per day.

Resources are important for the completion of any project on time as it is very important to see which task needs the resource. Resources need to be distributed in such a manner that the task doesn't suffer from any delay, and even if the task is delayed, then the delay is kept to a minimum and no priority tasks are delayed.

For example, consider a house building project. The white wash staff must be scheduled after the completion of in-wall piping and wall plastering. If they are scheduled before the completion of these two tasks, then the whitewash staff will face a delay, which will result in both cost and time wastage, and there might be a possibility that the whitewash staff might not be available for next slots or may charge more for the same. Therefore, all of these factors need to be kept in mind when resources are used.

Dependencies means that while the scheduling of the task is being done the task depends on the factors that are required for the completion or starting of the task (Jones 1997). Although dependencies can be classified into many forms, some of the important forms are categorized below:

- **Finish to Start (FS)**: To see if task A is finished so that the next consecutive task (task B) can be started.
- **Finish to Finish (FF)**: to see whether task A is completed.
- **Start to Start(SS)**: a task cannot be started until the previous task has not started.
- **Start to Finish(SF)**: task B cannot be completed until task A has started.

Budget along with the resource is an important aspect of completing a project because if project tasks are delayed then the budget for the project will also increase. Each task is assigned with resources such as time, money and staff, but if there is any variation in the resource of the task, all other factors will also suffer. Therefore, the main role of the scheduler is to keep a track on the task variance and act accordingly if any variance is noted in the completion of the task.

Project completion date is what all the project stakeholders want to see first. The completion of the project is a big milestone which is achieved after all the tasks are completed. All Gantt charts have the project completion date, and if there is any variance in the individual task, this date might be affected. This is one of the most important aspects of the Gantt chart as it is the final outcome date.

Milestones are somewhat similar to tasks but tasks have their duration constraints attached to them, whereas milestones have no such time parameter attached to them: they are either achieved or not achieved (Gray 1997).

We can categorize the milestone on the basis of the daily milestones or a task-wise milestone which are achieved or not achieved.

5.3.4 Evolutionary Algorithm in Developing a Software Project Management Methodology

We now work on the concept of retraining of the population or the data which is generated for obtaining the most optimized results from the population after inducing multiple mutated elements into the population.

The method of training the population set helps in calculating the project schedule. Managing the resources become more optimized as we have many new features added in the old method.

5.4 EXPECTED OUTPUTS OF THIS PROPOSED MODEL

- The scheduler designed for any project can be processed using this algorithm to obtain optimized and mutated parameterized results.
- The parameters need to be trained individually so that every parameter could have a self-generated and trained set of values, which can be used to ensure minimal chances of backdrop.
- This will not only give an optimized result but we can use it as an artificial scheduler, which can be used to schedule tasks with multiple mutated datasets.
- This optimized project planning timeline is an evolutionary scheduler which is developed at the end.

5.4.1 Business Implication

- Developing a scheduler like PERT, program evaluation and review technique chart, and Gantt chart with more risk considerations.
- Scheduling techniques developed are more optimized after using this model.
- All scheduling techniques can be mixed with computational intelligence techniques to obtain better results.
- The Gantt chart which we used earlier for defining the tasks, their implementation duration and when they will start processing is used for scheduling software engineering projects.
- This new evolutionary scheduler which is an optimized version of the Gantt chart is just evolving the previous dataset of similar models producing a more effective scheduler, which may be better and less prone to risks associated with the previous Gantt chart (Table 5.1).

TABLE 5.1
Comparison of Gantt Chart with the Evolutionary Scheduler

Gantt Chart	Evolutionary Scheduler
1. Scheduling is done using a project scheduler	Scheduling is done with the help of previously implemented scheduler datasets
2. Depends on the vision of the person who is assigning all the tasks and their required time	Takes the results from the existing scheduled models which will only process the best from these schedulers and generate an optimized result
3. Depends more on present conditions and less on future risks and aspects	Keeps a watch on the possible risks as its results are mutated from the existing schedulers
4. Generates results that may suffer from risks and less future prospects	Generates results on the calculations of risks and future aspects

5.5 CONCLUSION AND FUTURE SCOPE

It is clear that the evolutionary algorithm produces results which are definitely more optimized as the individual task undergoes the evolutionary process. This model will definitely impact the way the scheduler should be developed, so that the project is scheduled in a more compact manner. By using this model we can easily reduce the risk of any unexpected condition, as the results generated are generated by undergoing the mutation process. This model is considered to be a hybrid of the Gantt chart and evolutionary algorithm. Multiple scheduling techniques can be implemented using computational intelligence techniques. This will not only reduce the implementation time but will also reduce risk factors which delay the project, resulting in wastage of project resources, such as money, time and effort.

REFERENCES

Advanced Software Engineering Limited. Online from https://www.advsofteng.com/doc/cdcomdoc/gantt.htm. Retrieved on 27 March 2020.

Azar A.T. 2010. Adaptive neuro-fuzzy systems. In: Azar A.T. (ed) *Fuzzy Systems*. Croatia: InTech, pp. 85–105.

Conger, S. 2008. *The New Software Engineering*. Global Text Project. ISBN/ASIN: 0534171435. Zurich, Switzerland.

Duch W. 2007. What is computational Intelligence and Where Is It Going? In: Duch W., Mańdziuk J. (eds) *Challenges for Computational Intelligence. Studies in Computational Intelligence*. Vol. 63. Berlin, Heidelberg: Springer

Gray, A.R., MacDonell, S.G. 1997. A comparison of techniques for developing predictive models of software metrics. *Information and Software Technology*. 39, 6, pp. 425–437. ISSN 0950-5849. doi: 10.1016/S0950-5849(96)00006-7.

Hagan, M.T., Demuth, H.B., Beale, M.H.. 1996. *Neural Network Design*. Boston, MA: PWS Publishing.

Huges, B., Cotternel, M. 1999. *Software Project Management*, 2nd Ed. London: Tata McGraw Hill Education.

Humphrey, W. 1995. *A Discipline for Software Engineering*. Boston, MA: Addison Wesley Longman Publishing Co.

Huo S, He Z, Su J, Xi B, Zhu C. 2013. Using artificial neural network models for eutrophication prediction. *Procedia Environ Sci*. 18, pp. 310–316. doi: 10.1016/j.proenv.2013.04.040.

Jones, C. 1997. *Applied Software Measurement, Assuring Productivity and Quality*. New York: McGraw-Hill. ISBN-10: 0070328269

Konar A. 2005. *Computational Intelligence: Principles, Techniques and Applications*. Berlin: Springer.

Krasnopolsky V.M. 2013. *The Application of Neural Networks in the Earth System Sciences*. The Netherlands: Springer.

Leung, H., Fan, Z. 2002. *Handbook of Software Engineering and Knowledge Engineering*. Pittsburgh, PA: World Scientific.

Liou, Y.I. 1990. Knowledge acquisition: issues, techniques, and methodology. In *Proceedings of the 1990 ACM SIGBDP conference on Trends and directions in expert systems (SIGBDP '90)*. Association for Computing Machinery, New York, NY, USA, 212–236. https://doi.org/10.1145/97709.97726

Madani K. 2011. *Computational Intelligence*. Berlin: Springer.

Mukhopadhyay, T., Vicinanza, S.S. and Prietula, M.J. 1992. Examining the feasibility of a case-based reasoning model for software effort estimation. *MIS Quarterly*, 16: 155–171.

Neves A.C., González I., Leander J., Karoumi R. 2018. A new approach to damage detection in bridges using machine learning. In: Conte J., Astroza R., Benzoni G., Feltrin G., Loh K., Moaveni B. (eds) *Experimental Vibration Analysis for Civil Structures*. DOI: https://doi.org/10.1007/978-3-319-67443-8_5

Pandian, C.R. 2004. *Software Metrics: A Guide to planning, Analysis and Application*. Boca Raton, FL: Auerbach Publications.

Pressman, Roger S. 2005. *Software Engineering, A Practitioner's Approach Sixth Edition*. New York: McGraw-Hill.

Zhu, X. 2014. *Computational Intelligence Techniques and Applications*. Dordrecht: Springer.

6 Application of Intuitionistic Fuzzy Similarity Measures in Strategic Decision-Making

Anshu Ohlan
Pt. Neki Ram Sharma Government College

CONTENTS

6.1 INTRODUCTION

Computational intelligence techniques have found wide applications in solving the different types of problems of software engineering. The use of this field of artificial intelligence to the problems of software engineering is mainly in the form of decision support tools. These techniques are helpful in developing better quality software systems by incorporating uncertainty into the description of the system. The development in the field of fuzzy sets (FSs) theory has been found very helpful in dealing with practical situations of uncertain information. In fact, the notion of FSs and systems was propounded by the famous computer engineer Zadeh (1965) in his seminal paper "fuzzy sets." This has contributed toward a paradigm shift in almost all fields of science to allow things as a matter of degree. Such sets have proven to be very useful in the development of the field of computational intelligence (Bien et al. 2002; Malhotra and Bansal 2015). Because of their ability to efficiently deal with situations of uncertainty, these sets have been gaining remarkable recognition in different aspects of artificial intelligence (Dubois and Prade 1991). Zadeh (1968) proposed the notion of entropy as a technique for measuring fuzziness. The scientific literature on applications of fuzzy information measures to the computational intelligence field has been considerably extended by several researchers (Ohlan and Ohlan 2016). For instance, Atanassov (1986) pioneered the idea of Atanassov's intuitionistic fuzzy sets (AIFSs). Intuitionistic fuzzy sets (IFSs) have been used widely to solve problems concerning vagueness and lack of complete knowledge. The notion of IFSs can easily be used to introduce fuzzy sets (FSs) in situations where available information is insufficient to define vague concepts. Clearly, the notion of AIFSs is the extended version of FSs. The idea of vague sets was offered by Gau and Buehrer (1993). However, equivalency between vague sets and AIFSs was established by Bustince and Burillo (1996). The analysis of measures of similarity between IFSs has gained considerable scholarly attention in recent years. Inspired by the aforementioned development in FSs and systems, this chapter applies an exponential approach to IFSs. In this process, it introduces three reasonable measures to calculate similarity between two IFSs.

The advancement in computational technology has become an important feature of the solutions for modern multicriteria decision-making problems (Abdullah 2013). The remarkable progress in FSs theory has led to the development of many innovative tools for deriving optimal solutions for problems of decision science caused by uncertainty (Horiuchi et al. 2020; Yao 2020). In fact, fuzzy logic is a precise logic used in artificial intelligence (Zadeh 2008) and incorporates imprecise human perceptions. In this manner, it enhances the applications of artificial intelligence in providing cutting-edge solutions to problems of practical decision-making in cases of imperfect information (Ohlan 2015). Indeed, fuzzy logic is central to information processing. The fuzzy logic-based computational intelligence systems provide better solutions to multicriteria decision-making problems. Uncertainty is inevitable in various stages of software engineering. Because of the ability of fuzzy logic to efficiently deal with situations of vagueness, its use improves the operations of different software engineering tasks, such as incorporation of linguistic variables in the requirement phase, software testing and project management (Ohlan 2016). In addition, fuzzy logic is useful in the development of unsupervised learning, cyber-physical systems and many more. In sum, fuzzy systems allow the inclusion of human cognitive abilities in computing problems.

This chapter is arranged into eight sections. An analytical survey of relevant literature is presented in Section 6.2. Select widely used definitions related to theory of FSs and IFSs are given in Section 6.3. The novel methods of measuring similarity between IFSs are discussed in Section 6.4. This section also provides the evidence of their validity and applications in pattern recognition. Section 6.5 provides the application of our new exponential intuitionistic fuzzy similarity measures in strategic decision-making. The pros and cons of the introduced measures are discussed in Section 6.6. The next section summarizes the chapter. The scope for future research for applications of computational intelligence techniques to software engineering problems is discussed in the final section.

6.2 LITERATURE SURVEY

Because of their wide applicability in different fields, IFSs have witnessed significant expansion during recent decades. Primarily, the axiomatic definition of measures of similarity between specific IFSs was introduced by Dengfeng and Chuntian (2002). Liang and Shi (2003) reviewed a number of counterintuitive cases organizing from the tools intended by Dengfeng and Chuntian (2002). They defined numerous effective tools to measure similarity between IFSs in counterintuitive cases. Moreover, Mitchell (2003) presented counterintuitive outcome of Dengfeng and Chuntian (2002)'s similarity measure and provided a modified similarity measure with its application in pattern recognition problems. A new method of computing distance between IFSs employing Hausdorff distance was introduced by Hung and Yang (2004). Thereafter, a number of associations between similarity and distance measure of IFSs were investigated by Wang and Xin (2005). Liu (2005) presented the unreasonable cases of the similarity measures of select extant studies: Chen (1995, 1997), Hong and Kim (1999) and Dengfeng and Chuntian (2002). They also introduced several new IF measures of similarity. The select different measures of similarity between IFSs were introduced by Zhang and Fu (2006), Park et al. (2007), Hung and Yang (2007), Vlachos and Sergiadis (2007), Xu (2007), Hung and Yang (2008a, 2008b), Ye (2011), Li et al. (2012), Julian et al. (2012), Hwang and Yang (2012), Boran and Akay (2014), Tan and Chen (2014), Farhadinia (2014), Chen and Chang (2015), Chen et al. (2016), Mishra et al. (2017), Wei and Wei (2018), and Garg and Kumar (2018).

The existing measures of similarity between IFSs have considerable significance in the field of artificial intelligence. However, there is a good scope for introducing more efficient similarity measures to deal with situations of fuzziness in the development of time-efficient computational intelligence techniques (Agrawal et al. 2019). The exponential IF similarity measures introduced in the current work may be efficiently used to solve software engineering problems.

6.3 PROBLEM IDENTIFICATION

We now explain select notions connected with the theoretical aspects of FSs and IFSs.

Definition 1. ***Fuzzy Set*** (Zadeh 1965): It is assumed that $V = (v_1, v_2, ..., v_n)$ is the given finite nonempty set. A FS E on V may be stated as:

$$E = \{\langle v, \mu_E(v)\rangle / v \in V\} \tag{6.1}$$

where $\mu_E : V \rightarrow [0,1]$ is mapping of membership of E and $\mu_E\ (v)$ demontrates the amount of the belongingness of $v \in V$ in E.

Definition 2. ***Intuitionistic Fuzzy Set** (IFS)* (Atanasov 1986): Assuming $V = (v_1, v_2, \ldots, v_n)$ is the given finite nonempty set. An IFS I on V is expressed as:

$$I = \{\langle v_i, \mu_I(v_i), \nu_I(v_i)\rangle / v_i \in V\} \tag{6.2}$$

where $\mu_I : V \to [0,1]$, $\nu_I : V \to [0,1]$ with the condition $0 \le \mu_I(v_i) + \nu_I(v_i) \le 1$, $\forall v_i \in V$. $\mu_I(v_i)$, $\nu_I(v_i) \in [0,1]$ signifies the amount of membership and nonmembership of v_i in I, correspondingly.

For an IFS I, the term $\pi_I(v_i) = 1 - \mu_I(v_i) - \nu_I(v_i)$ is the amount of hesitancy of v_i in I and $0 \le \pi_I(v_i) \le 1$ for every $v_i \in V$. For a FS I in V, $\pi_I(v_i) = 0$ when $\nu_I(v_i) = 1 - \mu_I(v_i)$. Consequently, FS is found to be a particular case of IFS.

Hung and Yang (2004), Tan and Chen (2014) and Chen and Chang (2015) presented the axioms for validation of an IF similarity measure as follows:

i. $0 \le S, (D, P) \le 1$
ii. $S(D, P) = 1$ if and only if $D = P$.
iii. $S(D, P) = S(P, D)$
iv. If $D \subseteq P \subseteq J$, $D, P, J \in$ IFSs(Z)

Then $S(D, J) \le S(D, P)$ and $S(D, J) \le S(P, J)$.

6.4 EXPONENTIAL INTUITIONISTIC FUZZY MEASURES OF SIMILARITY

It is very prominent that the concepts of similarity and divergence are dual to each other. It is assumed that $V = (v_1, v_2, \ldots, v_n)$ be a finite universal set, and I_1 and I_2 are two IFSs comprising the membership and nonmembership values $\mu_{I_1}(v_i)$, $\mu_{I_2}(v_i)$ $\forall v_i \in V$. Here, we utilize the measure of divergence $D^E_{\text{IFS}}(I_1, I_2)$ established by Ohlan and Ohlan (2016) to introduce the new measures of similarity as:

$$S_E(I_1, I_2) = 1 - D^E_{\text{IFS}}(I_1, I_2) \tag{6.3}$$

where

$D^E_{\text{IFS}}(I_1, I_2)$

$$= \sum_{i=1}^{n} \left[\begin{array}{l} 2 - \left(1 - \dfrac{(\mu_{I_1}(v_i) - \mu_{I_2}(v_i)) - (\nu_{I_1}(v_i) - \nu_{I_2}(v_i))}{2}\right) e^{\left(\frac{(\mu_{I_1}(v_i) - \mu_{I_2}(v_i)) - (\nu_{I_1}(v_i) - \nu_{I_2}(v_i))}{2}\right)} \\ - \left(1 + \dfrac{(\mu_{I_1}(v_i) - \mu_{I_2}(v_i)) - (\nu_{I_1}(v_i) - \nu_{I_2}(v_i))}{2}\right) e^{\left(\frac{(\nu_{I_1}(v_i) - \nu_{I_2}(v_i)) - (\mu_{I_1}(v_i) - \mu_{I_2}(v_i))}{2}\right)} \end{array} \right]$$

Hence, we introduce another measure of similarity using the exponential operation as

$$S_e(I_1,I_2) = \frac{e^{-D^E_{\text{IFS}}(I_1,I_2)} - e^{-1}}{1-e^{-1}} \tag{6.4}$$

Again, we propose our next measure of similarity as

$$S_c(I_1,I_2) = \frac{1 - D^E_{\text{IFS}}(I_1,I_2)}{1 + D^E_{\text{IFS}}(I_1,I_2)} \tag{6.5}$$

Theorem 1. The proposed IFSs in Equations 6.3–6.5 are effective IF similarity measures.

Proof: Assuming a monotonic decreasing function g and in view of the fact that $0 \leq D^E_{\text{IFS}}(I_1,I_2) \leq 1$.

Accordingly, we have $g(1) \leq g\left(D^E_{\text{IFS}}(I_1,I_2)\right) \leq g(0)$, and for a monotonic decreasing function the remaining other axioms will also hold.

Now, it is required to choose a convenient function g for every situation.

i. Let us suppose a function g as $g_1(y) = 1 - y$. Figure 6.1 displays that function g_1 is a monotonic decreasing function.

   The similarity measure $S_E(I_1, I_2) = 1 - D^E_{\text{IFS}}(I_1, I_2)$ in Equation 6.3 is defined, and Figure 6.2 represents that $0 \leq S_E(I_1, I_2) \leq 1$

   Hence, $S_E(I_1, I_2)$ is a valid measure of similarity between IFSs.

ii. Now let us assume an exponential mapping as $g_2(y) = e^{-y}$. Figure 6.3 depicts that function g_2 is a monotonic decreasing function.

   Therefore, we have a measure in Equation 6.4, $S_e(I_1,I_2) = \frac{e^{-D^E_{\text{IFS}}(I_1,I_2)} - e^{-1}}{1-e^{-1}}$ is defined according to the definition given above.

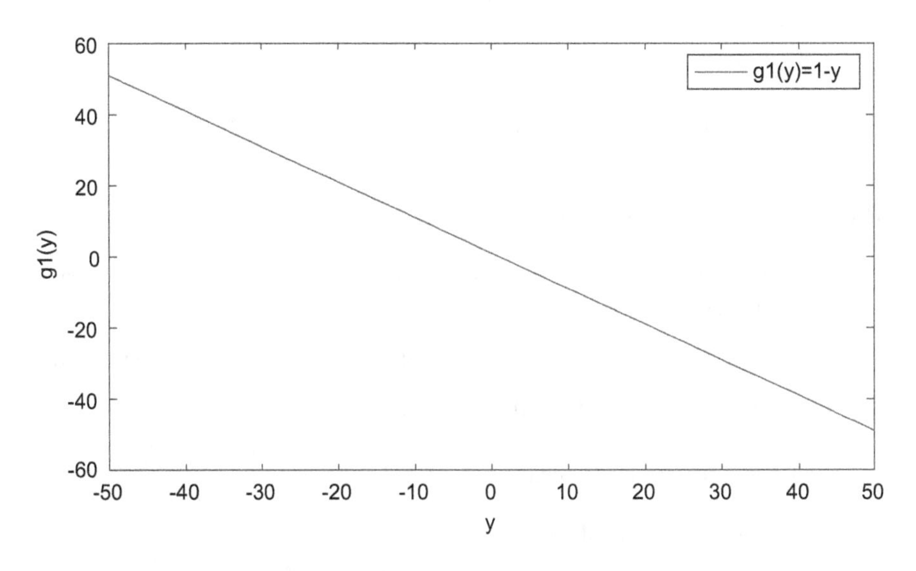

FIGURE 6.1 Behavior of function $g_1(y) = 1 - y$.

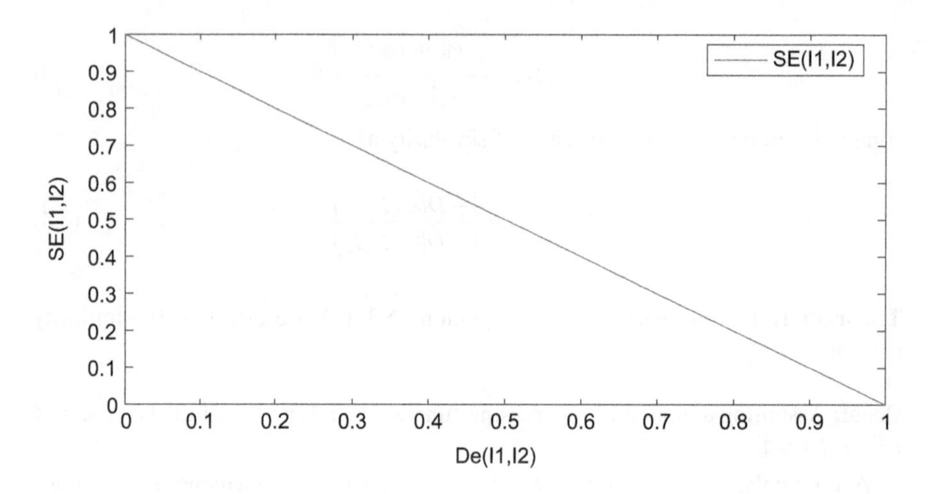

FIGURE 6.2 IF similarity measure $S_E(I_1, I_2)$.

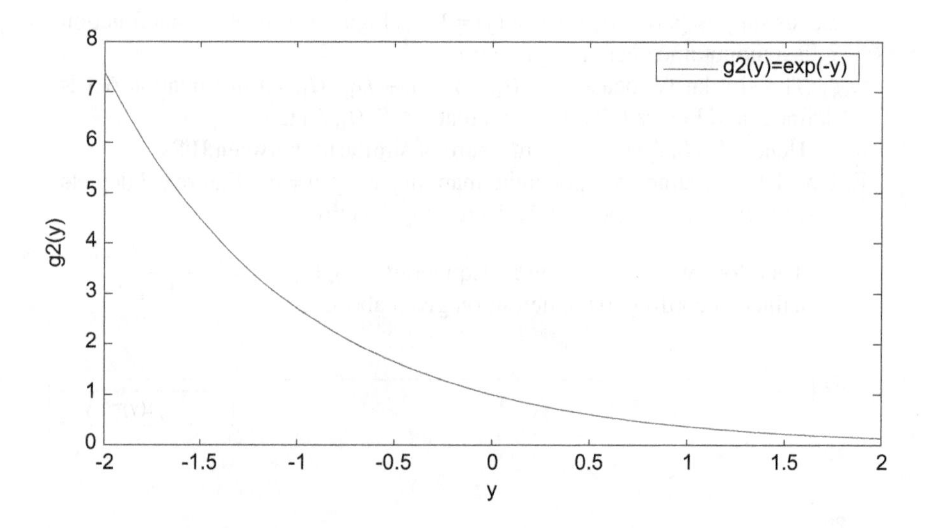

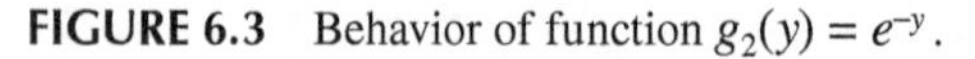

FIGURE 6.3 Behavior of function $g_2(y) = e^{-y}$.

Figure 6.4 signifies the nature of IF similarity measure that $0 \le S_e(I_1, I_2) \le 1$.

iii. Further, if we select a mapping of g as $g_3(y) = \dfrac{1}{1+y}$, as can be seen from Figure 6.5, g_3 is a monotonically decreasing function.

We obtained the similarity measure (6.5) $S_c(I_1, I_2) = \dfrac{1 - D_{\text{IFS}}^E(I_1, I_2)}{1 + D_{\text{IFS}}^E(I_1, I_2)}$ in view of Figure 6.6, which is explained above.

This proves the validity of the IFSM proposed in Equations 6.3–6.5.

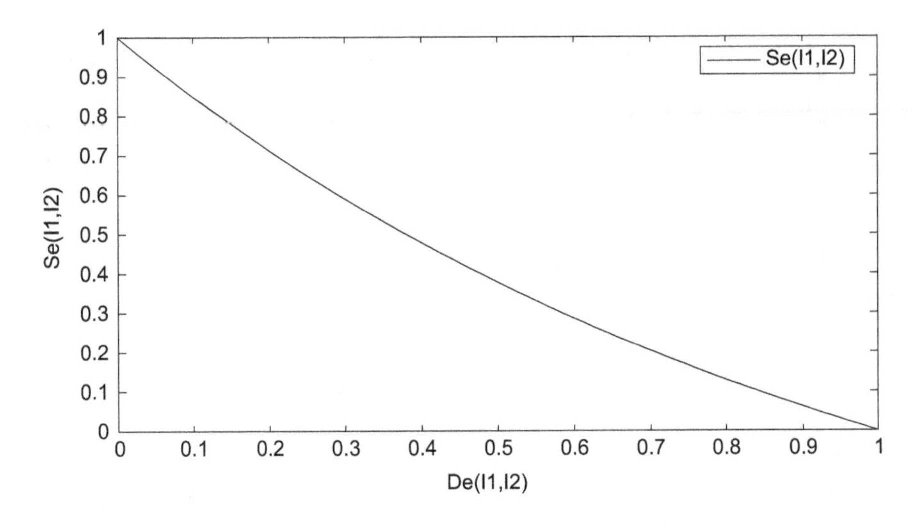

FIGURE 6.4 IF similarity measure $S_e(I_1, I_2)$.

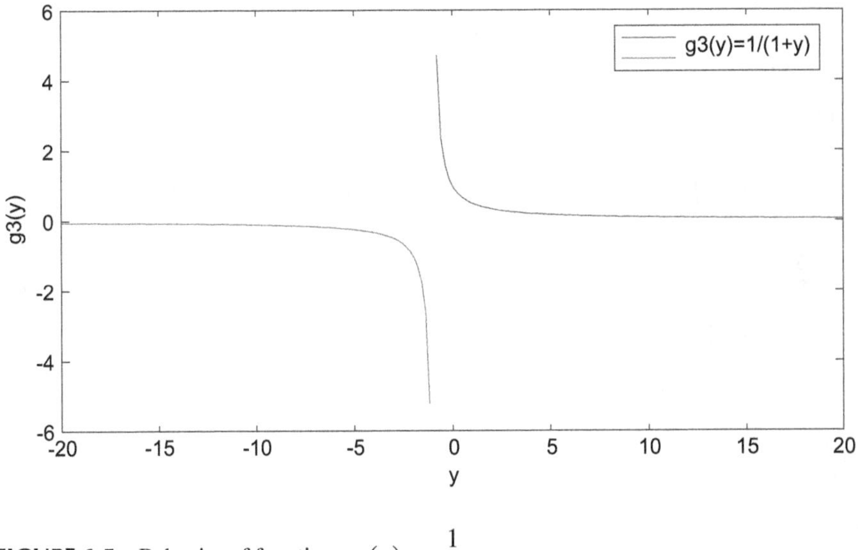

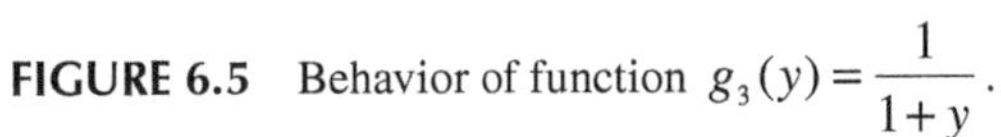

FIGURE 6.5 Behavior of function $g_3(y) = \dfrac{1}{1+y}$.

Definition 3. Given two IFSs I_1 and I_2 are defined in a discrete universal set $T = (t_1, t_2, ..., t_k)$ with membership values $\mu_{I_1}(t_s)$, $\mu_{I_2}(t_s)$ $\forall t_s \in T$ and with nonmembership values $v_{I_1}(t_s)$, $v_{I_2}(t_s)$ $\forall v_i \in V$. A weight vector $W = \langle W_1, W_2, ..., W_k \rangle'$ of elements t_s, $\forall s \in \{1,2,...,k\}$ such that $W_s \geq 0$ and $\sum_{s=1}^{k} W_s = 1$, the weighted exponential measure of similarity between IFSs I_1 and I_2 is given as

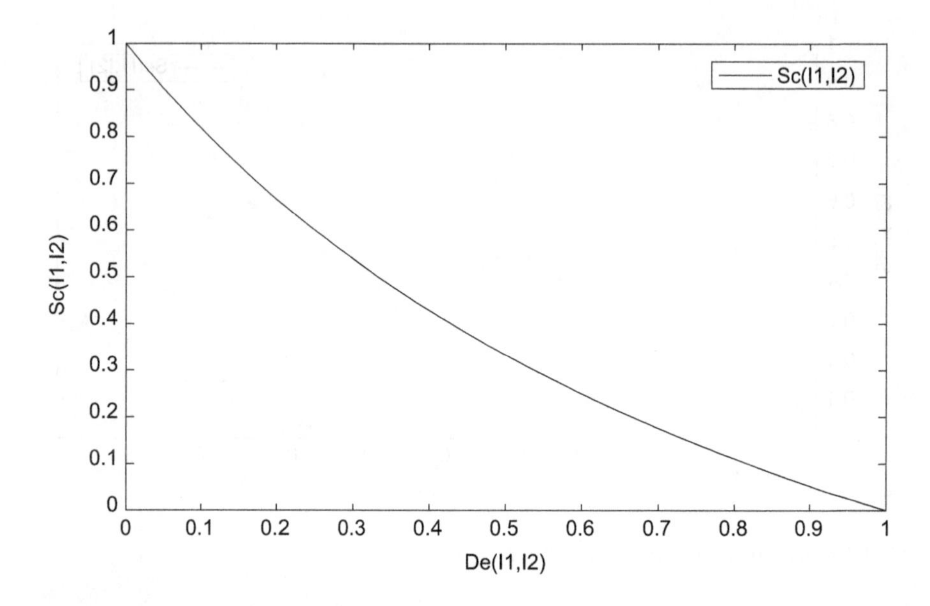

FIGURE 6.6 IF similarity measure $S_C(I_1,I_2)$.

i.

$$S_E^W(I_1,I_2) = 1-\sum_{s=1}^{k} W_s \left[2-\left(1-\frac{(\mu_{I_1}(t_s)-\mu_{I_2}(t_s))-(\nu_{I_1}(t_s)-\nu_{I_2}(t_s))}{2}\right)e^{\left(\frac{(\mu_{I1}(t_s)-\mu_{I2}(t_s))-(\nu_{I1}(t_s)-\nu_{I2}(t_s))}{2}\right)} -\left(1+\frac{(\mu_{I_1}(t_s)-\mu_{I_2}(t_s))-(\nu_{I_1}(t_s)-\nu_{I_2}(t_s))}{2}\right)e^{\left(\frac{(\nu_{I1}(t_s)-\nu_{I2}(t_s))-(\mu_{I1}(t_s)-\mu_{I2}(t_s))}{2}\right)} \right] \tag{6.6}$$

ii.

$$S_e^W(I_1,I_2) = \frac{\exp\left(-\sum_{s=1}^{k} W_s \left[2-\left(1-\frac{(\mu_{I_1}(t_s)-\mu_{I_2}(t_s))-(\nu_{I_1}(t_s)-\nu_{I_2}(t_s))}{2}\right)e^{\left(\frac{(\mu_{I1}(t_s)-\mu_{I2}(t_s))-(\nu_{I1}(t_s)-\nu_{I2}(t_s))}{2}\right)} -\left(1+\frac{(\mu_{I_1}(t_s)-\mu_{I_2}(t_s))-(\nu_{I_1}(t_s)-\nu_{I_2}(t_s))}{2}\right)e^{\left(\frac{(\nu_{I1}(t_s)-\nu_{I2}(t_s))-(\mu_{I1}(t_s)-\mu_{I2}(t_s))}{2}\right)} \right]\right)-\exp(-1)}{1-\exp(-1)} \tag{6.7}$$

iii.

$$S_c^W(I_1,I_2) = \frac{1-\left(\sum_{s=1}^{k} W_s \left[2-\left(1-\frac{(\mu_{I_1}(t_s)-\mu_{I_2}(t_s))-(\nu_{I_1}(t_s)-\nu_{I_2}(t_s))}{2}\right)e^{\left(\frac{(\mu_{I_1}(t_s)-\mu_{I_2}(t_s))-(\nu_{I_1}(t_s)-\nu_{I_2}(t_s))}{2}\right)} -\left(1+\frac{(\mu_{I_1}(t_s)-\mu_{I_2}(t_s))-(\nu_{I_1}(t_s)-\nu_{I_2}(t_s))}{2}\right)e^{\left(\frac{(\nu_{I_1}(t_s)-\nu_{I_2}(t_s))-(\mu_{I_1}(t_s)-\mu_{I_2}(t_s))}{2}\right)} \right]\right)}{1+\left(\sum_{s=1}^{k} W_s \left[2-\left(1-\frac{(\mu_{I_1}(t_s)-\mu_{I_2}(t_s))-(\nu_{I_1}(t_s)-\nu_{I_2}(t_s))}{2}\right)e^{\left(\frac{(\mu_{I_1}(t_s)-\mu_{I_2}(t_s))-(\nu_{I_1}(t_s)-\nu_{I_2}(t_s))}{2}\right)} -\left(1+\frac{(\mu_{I_1}(t_s)-\mu_{I_2}(t_s))-(\nu_{I_1}(t_s)-\nu_{I_2}(t_s))}{2}\right)e^{\left(\frac{(\nu_{I_1}(t_s)-\nu_{I_2}(t_s))-(\mu_{I_1}(t_s)-\mu_{I_2}(t_s))}{2}\right)} \right]\right)} \quad (6.8)$$

It is observed that if $W=\left(\frac{1}{k},\frac{1}{k},\frac{1}{k},\ldots,\frac{1}{k}\right)$, the weighted exponential IF similarity measures (6.6) to (6.8) are reduced to (6.3) to (6.5).

In the next section, we present a method to resolve the problems concerning business decision-making by means of weighted exponential IF similarity measures (6.6) to (6.8).

6.4.1 Effectiveness of the Offered Measures of Similarity in Pattern Recognition

We now exhibit the competency of the proposed IF measures of similarity in pattern recognition by comparing them with the extant IF measures of similarity presented in Chen (1995), Hong and Kin (1999), Li and Xu (2001), Ye (2011) and Song et al. (2015).

Let J_I and J_F be two IF subsets of a finite set $S = \{s_1, s_2, \ldots s_j\}$.

Chen (1995) introduced the measure of similarity between IFSs J_I and J_F as

$$S_C(J_I,J_F)=1-\frac{\sum_{p=1}^{j}\left|(\mu_{J_I}(s_p)-\nu_{J_I}(s_p))-(\mu_{J_F}(s_p)-\nu_{J_F}(s_p))\right|}{2j} \quad (6.9)$$

Hong and Kin (1999) defined the IF similarity measure between J_I and J_F as

$$S_{HK}(J_I,J_F)=1-\frac{\sum_{p=1}^{j}\left(\left|\mu_{J_I}(s_p)-\mu_{J_F}(s_p)\right|+\left|(\nu_{J_I}(s_p)-\nu_{J_F}(s_p)\right|\right)}{2j} \quad (6.10)$$

Li and Xu (2001) provided the IF similarity measure in IFSs J_I and J_F given by

$$S_{LX}(J_I, J_F) = 1 - \frac{\sum_{p=1}^{j} \left| \left(\mu_{J_I}(s_p) - v_{J_I}(s_p)\right) - \left(\mu_{J_F}(s_p) - v_{J_F}(s_p)\right) \right|}{4j} - \frac{\sum_{p=1}^{j} \left(\left|\mu_{J_I}(s_p) - \mu_{J_F}(s_p)\right| + \left|v_{J_I}(s_p) - v_{J_F}(s_p)\right| \right)}{4j} \tag{6.11}$$

In 2011, Ye (2011) presented the following similarity measure in IF situation as

$$C_{YeIFS}(J_I, J_F) = \frac{1}{j} \sum_{p=1}^{j} \frac{\mu_{J_I}(s_p)\mu_{J_F}(s_p) + v_{J_I}(s_p) v_{J_F}(s_p)}{\sqrt{\left(\mu_{J_I}(s_p)\right)^2 + \left(v_{J_I}(s_p)\right)^2} \sqrt{\left(\mu_{J_F}(s_p)\right)^2 + \left(v_{J_F}(s_p)\right)^2}} \tag{6.12}$$

Song et al. (2015) established the IF similarity measure as

$$S_Y(J_I, J_F) = \frac{1}{2j} \sum_{p=1}^{j} \left(\begin{array}{c} \sqrt{\mu_{J_I}(s_p)\mu_{J_F}(s_p)} + 2\sqrt{v_{J_I}(s_p) v_{J_F}(s_p)} + \sqrt{\pi_{J_I}(s_p)\pi_{J_F}(s_p)} \\ + \sqrt{\left(1 - v_{J_I}(s_p)\right)\left(1 - v_{J_F}(s_p)\right)} \end{array} \right) \tag{6.13}$$

To solve the problems of pattern recognition in IF scenario we use the following technique.

Assuming we are given c identified patterns $\breve{P}_{I1}, \breve{P}_{I2}, \breve{P}_{I3}, \ldots, \breve{P}_{Ic}$ having classifications as , $\breve{C}_{I1}, \breve{C}_{I2}, \breve{C}_{I3}, \ldots, \breve{C}_{Ic}$, respectively. $S = \{s_1, s_2, \ldots s_j\}$ is assumed to be the finite set. The patterns are signified by the subsequent IFSs as $\breve{P}_{Il} = \left\{ \left\langle s_l, \mu_{\breve{P}_{Il}}(s_l), v_{\breve{P}_{Il}}(s_l) \right\rangle / s_l \in S \right\}$, where $l = 1, 2, \ldots, c$ and $k = 1, 2, \ldots, j$.

If $\breve{Q}_F$ is an unidentified pattern characterized by an IFS

$$\breve{Q}_{Fl} = \left\{ \left\langle s_l, \mu_{\breve{Q}_{Fl}}(s_l), v_{\breve{Q}_{Fl}}(s_l) \right\rangle / s_l \in S \right\}.$$

Our target is to classify $\breve{Q}_{Fl}$ to one of the classes $\breve{C}_{I1}, \breve{C}_{I2}, \breve{C}_{I3}, \ldots, \breve{C}_{Ic}$. In line with the rule of having the maximum similarity between IFSs, the process of assigning $\breve{Q}_{Fl}$ to $\breve{C}_{Iz*}$ is expressed as

$$Iz^* = \arg\max_{Iz} \left\{ S\left(P_{Iz}, \breve{Q}_{Fl}\right) \right\}.$$

According to the above algorithm, a pattern may be acknowledged to select the most excellent category.

TABLE 6.1
Computed Numerical Values of IF Similarity Measures

$\breve{Q}_{Fl}$	$\breve{P}_{I1}$	$\breve{P}_{I2}$	$\breve{P}_{I3}$	$\breve{P}_{I4}$
$S_C(\breve{P}_{Il}, \breve{Q}_{Fl})$	0.7667	0.8500	0.8167	0.7833
$S_{HK}(\breve{P}_{Il}, \breve{Q}_{Fl})$	0.7333	0.7833	0.7833	0.7500
$S_{LX}(\breve{P}_{Il}, \breve{Q}_{Fl})$	0.7500	0.8167	0.8000	0.7667
$C_{YeIFS}(\breve{P}_{Il}, \breve{Q}_{Fl})$	0.8724	0.9153	0.9145	0.8694
$S_Y(\breve{P}_{Il}, \breve{Q}_{Fl})$	0.9060	0.9444	0.9222	0.9074
$S_E(\breve{P}_{Il}, \breve{Q}_{Fl})$	0.7063	0.8345	0.8285	0.7390
$S_e(\breve{P}_{Il}, \breve{Q}_{Fl})$	0.5974	0.7587	0.7507	0.6366
$S_c(\breve{P}_{Il}, \breve{Q}_{Fl})$	0.5460	0.7160	0.7072	0.5860

Example 1. It is assumed that there are three recognized patterns $\breve{P}_{I1}$, $\breve{P}_{I2}$, $\breve{P}_{I3}$ and $\breve{P}_{I4}$ having the classifications $\breve{C}_{I1}$, $\breve{C}_{I2}$ and $\breve{C}_{I3}$, respectively. The following IFSs represent the given three known patterns with the set $S = \{s_1, s_2, s_3\}$:

$$\breve{P}_{I1} = \{\langle s_1, 0.9, 0.1\rangle, \langle s_2, 0.7, 0.1\rangle, \langle s_3, 0.7, 0.2\rangle\},$$

$$\breve{P}_{I2} = \{\langle s_2, 0.7, 0.2\rangle, \langle s_2, 0.9, 0.1\rangle, \langle s_3, 0.8, 0.1\rangle\},$$

$$\breve{P}_{I3} = \{\langle s_1, 0.5, 0.3\rangle, \langle s_2, 0.7, 0.1\rangle, \langle s_3, 0.5, 0.3\rangle\},$$

$$\breve{P}_{I4} = \{\langle s_1, 0.8, 0.1\rangle, \langle s_2, 0.5, 0.3\rangle, \langle s_3, 0.6, 0.3\rangle\},$$

where $l = 1,2,3,4$ represents

$$P_{Il} = \{\langle s_1, \mu_{P_{Il}}(s_1), \nu_{P_{Il}}(s_1)\rangle, \langle s_2, \mu_{P_{Il}}(s_1), \nu_{P_{Il}}(s_1)\rangle, \langle s_3, \mu_{P_{Il}}(s_1), \nu_{P_{Il}}(s_1)\rangle\}.$$

We have an unknown pattern $\breve{Q}_{Fl}$, which is characterized by IFS

$$\breve{Q}_{Fl} = \{\langle s_1, 0.4, 0.4\rangle, \langle s_2, 0.6, 0.3\rangle, \langle s_3, 0.9, 0.0\rangle\}.$$

Here, we intend to categorize $\breve{Q}_{Fl}$ to one of the pattern $\breve{P}_{I1}$, $\breve{P}_{I2}$, $\breve{P}_{I3}$ and $\breve{P}_{I4}$ having the classifications $\breve{C}_{I1}$, $\breve{C}_{I2}$, $\breve{C}_{I3}$ and $\breve{C}_{I4}$ respectively.

Using the above algorithm, we compute the values of different IF similarity measures in Equations 6.3–6.5 and 6.9–6.15. The computed values are shown in Table 6.1.

From the obtained quantitative results of numerous extant measures of similarity and the measures proposed here S_E, S_e, S_c (listed in Table 6.1), it is clear that the pattern $\breve{Q}_{Fl}$ is the most similar with the pattern $\breve{P}_{I2}$. However from Row 2, it is also observed that the measure S_{HK} is classifying the pattern $\breve{Q}_{Fl}$ to $\breve{P}_{I2}$ and $\breve{P}_{I3}$. Thus, the proposed IF similarity measures are effective and efficient for pattern recognition applications.

6.5 APPLICATION OF WEIGHTED EXPONENTIAL INTUITIONISTIC FUZZY SIMILARITY MEASURES

Vagueness is inevitably involved in decisions in the corporate world. The index of fuzziness used by Zadeh to introduce the theory of FSs considers a single degree of belongingness from a function. However, complex management situations necessitate paying attention to degrees of both belongingness and absence to solve practical business and decision problems. Hence, the advancement in fuzzy science has played a pivotal role in dealing with challenges arising as a result of situations involving imperfect information (Szmidt 2014). The generalizations of FSs such as IFSs can be seen as efficient tools to express imprecision in computational algorithms and processes. IFSs have the capability to incorporate the third state of fuzziness: neither here nor there. Therefore, decision-making with IFSs is an extension of fuzzy decisions. The weighted exponential IF measures of similarity have become an essential part of strategic business decision-making. Clearly, the measures of similarities between IFSs are highly useful in cases of business decision-making where available information is not sufficient to define imprecise concepts (Li 2014).

6.5.1 Computational Approach

Here, we discuss the application of the newly introduced exponential IF measures of similarity in a framework of multiattribute decision-making. Let's assume that a set of m alternatives be $O = \{O_1, O_2, O_3, \ldots, O_m\}$ and an attribute set $N = \{N_1, N_2, N_3, \ldots, N_n\}$. The deciding authority has to opt the best one from set O in accordance with the attribute set N.

We introduce a technique to respond to the decision-making scenario using the measures (6.3) to (6.5); the computational approach of the process is as follows:

1. Construct an IF decision matrix (IFDM)
   The IFDM $M = (m_{ji})_{m \times n}$ of IF value $m_{ji} = \{\mu_{ji}, \nu_{ji}\}$ is constructed.
2. Establish the IF positive ideal solution (IFPIS) and IF negative ideal solution (IFNIS)

$$\text{IFPIS} = O^{+} = \{\langle \mu_1^{+}, \nu_1^{+} \rangle, \langle \mu_2^{+}, \nu_2^{+} \rangle, \ldots, \langle \mu_n^{+}, \nu_n^{+} \rangle\} \tag{6.9}$$

$$\text{IFNIS} = O^{-} = \{\langle \mu_1^{-}, \nu_1^{-} \rangle, \langle \mu_2^{-}, \nu_2^{-} \rangle, \ldots, \langle \mu_n^{-}, \nu_n^{-} \rangle\} \tag{6.10}$$

3. Calculate $S^{W^+} = S^W(O^+, O_j)$ and $S^{W^-} = S^W(O^-, O_j)$, $j = 1, 2, \ldots, m$ using the weighted similarity measures (6.6) to (6.8).
4. Calculate comparative similarity of the optimal solution.
   The comparative similarity of alternative O_j with regard to IFPIS and IFNIS is given by

$$S^W = \frac{S^{W^+}}{S^{W^-} + S^{W^+}}, \; j = 1, 2, \ldots, m. \tag{6.11}$$

5. Select the best option O_j in line with the largest S_j.

The above method is demonstrated with the help of examples below.

Example 2. As discussed in Wei (2008) and Wei et al. (2011), an investment corporation wishes to devote an assured sum of funds in the topmost alternative among the following five options: H_1, a car showroom; H_2, a fast food point; H_3, a computer solution hub; H_4, a business of library solution, and H_5, a business of air conditioners. This corporation wishes to make decisions in accordance with the following four characteristics: (1) the maximization of socioeconomic impact K_1, (2) the maximization of growth K_2, (3) the minimization of risk K_3 and (4) the minimization of environmental consequences K_4, having weight considerations $W = \langle 0.32, 0.18, 0.20, 0.30\rangle'$. The deciding authority has to select the most excellent H_j according to the characteristics K_i, i = 1, 2, 3, 4.

Step 1: An IFDM $M = (m_{ji})_{5\times4}$ is presented in Table 6.2.
Step 2: The IFPIS and IFNIS are obtained using the formulae (6.9) and (6.10), respectively. Table 6.3 lists the results.
Step 3: Tables 6.4 and 6.5 display the computed values of $S^{W^+} = S^W(H^+,H_j)$ and $S^{W^-} = S^W(H^-,H_j)$ using the weighted similarity measures (6.6) to (6.8).

TABLE 6.2
IF Decision Matrix $M = (m_{ji})_{5\times4}$

	K_1	K_2	K_3	K_4
H_1	{0.3, 0.6}	{0.6, 0.3}	{0.5, 0.4}	{0.2, 0.7}
H_2	{0.7, 0.2}	{0.7, 0.2}	{0.7, 0.3}	{0.4, 0.5}
H_3	{0.5, 0.3}	{0.5, 0.4}	{0.6, 0.4}	{0.6, 0.3}
H_4	{0.3, 0.4}	{0.6, 0.3}	{0.8, 0.1}	{0.2, 0.6}
H_5	{0.7, 0.1}	{0.4, 0.3}	{0.6, 0.2}	{0.5, 0.3}

TABLE 6.3
IFPIS H^+ and IFNIS H^-

	K_1	K_2	K_3	K_4
H^+	{0.7, 0.1}	{0.7, 0.2}	{0.8, 0.1}	{0.6, 0.3}
H^-	{0.3, 0.6}	{0.4, 0.4}	{0.5, 0.4}	{0.2, 0.7}

TABLE 6.4
Values of Weighted Similarity Measures $S^W(H^+, H_j)$

$S_E^W(H^+, H_1)$	0.8621	$S_e^W(H^+, H_1)$	0.7999	$S_c^W(H^+, H_1)$	0.7652
$S_E^W(H^+, H_2)$	0.9826	$S_e^W(H^+, H_2)$	0.9729	$S_c^W(H^+, H_2)$	0.9663
$S_E^W(H^+, H_3)$	0.9671	$S_e^W(H^+, H_3)$	0.9492	$S_c^W(H^+, H_3)$	0.9373
$S_E^W(H^+, H_4)$	0.9199	$S_e^W(H^+, H_4)$	0.8808	$S_c^W(H^+, H_4)$	0.8574
$S_E^W(H^+, H_5)$	0.9875	$S_e^W(H^+, H_5)$	0.9686	$S_c^W(H^+, H_5)$	0.9757

TABLE 6.5
Values of Weighted Similarity Measures $S^W(H^-, H_j)$

$S_E^W(H^-, H_1)$	0.9959	$S_e^W(H^-, H_1)$	0.9936	$S_c^W(H^-, H_1)$	0.9920
$S_E^W(H^-, H_2)$	0.9186	$S_e^W(H^-, H_2)$	0.8790	$S_c^W(H^-, H_2)$	0.8550
$S_E^W(H^-, H_3)$	0.9288	$S_e^W(H^-, H_3)$	0.8950	$S_c^W(H^-, H_3)$	0.8743
$S_E^W(H^-, H_4)$	0.9736	$S_e^W(H^-, H_4)$	0.9596	$S_c^W(H^-, H_4)$	0.9505
$S_E^W(H^-, H_5)$	0.8890	$S_e^W(H^-, H_5)$	0.8388	$S_c^W(H^-, H_5)$	0.8106

TABLE 6.6
Results of $S^W(O_j)$

$S_E^W(H_1)$	0.4640	$S_e^W(H_1)$	0.4460	$S_c^W(H_1)$	0.4355
$S_E^W(H_2)$	0.5168	$S_e^W(H_2)$	0.5254	$S_c^W(H_2)$	0.5306
$S_E^W(H_3)$	0.5101	$S_e^W(H_3)$	0.5147	$S_c^W(H_3)$	0.5174
$S_E^W(H_4)$	0.4858	$S_e^W(H_4)$	0.4786	$S_c^W(H_4)$	0.4743
$S_E^W(H_5)$	0.5262	$S_e^W(H_5)$	0.5359	$S_c^W(H_5)$	0.5462

Step 4: The calculated numerical values of the comparative similarity of the ideal solutions employing the formula (6.11) are reported in Table 6.6.

Step 5: Thus, H_5 is the most preferable alternative that matched with the result derived in Wei (2008) and Wei et al. (2011).

Example 3. As given in Li (2005), consider an air conditioner selection problem in which a person has three options B_j, $j = 1,2,3$ of choosing an air conditioner available in the market with respect to three qualities N_1 (economical), N_2 (functional), N_3 (being operative). Assume weight vector as $W = \langle 0.25, 0.45, 0.30 \rangle'$. The decision-maker analyzes the alternatives B_j, $j = 1,2,3$ in the context of attributes N_i, $i = 1,2,3$.

Step 1: The IFDM $M = (m_{ji})_{3\times3}$ is presented in Table 6.7.

Step 2: The IFPIS and IFNIS are computed employing formulae (6.9) and (6.10), respectively. Table 6.8 shows the obtained results.

Step 3: Tables 6.9 and 6.10 show the computed values of $S^{W^+} = S^W(B^+,B_j)$ and $S^{W^-} = S^W(B^-,B_j)$ using the weighted similarity measures (6.6) to (6.8).

Step 4: Calculated numerical values of the comparative similarity of the ideal solution applying formula (6.11) are shown in Table 6.11.

Step 5: Thus, B_1 is the most preferable alternative which matched with the result obtained in Li (2005).

6.6 PROS AND CONS OF THE OFFERED SOLUTIONS

The science of software engineering strives to obtain reliable and efficient solutions for optimal allocation of scarce resources. The basic use of computational

TABLE 6.7
IFDM $M = (m_{ji})_{3\times3}$

	N_1	N_2	N_3
B_1	{0.75, 0.10}	{0.60, 0.25}	{0.80, 0.20}
B_2	{0.80, 0.15}	{0.68, 0.20}	{0.45, 0.50}
B_3	{0.40, 0.45}	{0.75, 0.05}	{0.60, 0.30}

TABLE 6.8
IFPIS B^+ and IFNIS B^-

	N_1	N_2	N_3
B^+	{0.80, 0.10}	{0.75, 0.05}	{0.80, 0.20}
B^-	{0.40, 0.45}	{0.60, 0.25}	{0.45, 0.50}

TABLE 6.9
Results of Weighted Similarity Measures $S^W(B^+, B_j)$

$S_E^W(B^+, B_1)$	0.9859	$S_e^W(B^+, B_1)$	0.9781	$S_c^W(B^+, B_1)$	0.9727
$S_E^W(B^+, B_2)$	0.9619	$S_e^W(B^+, B_2)$	0.9424	$S_c^W(B^+, B_2)$	0.9302
$S_E^W(B^+, B_3)$	0.9568	$S_e^W(B^+, B_3)$	0.9358	$S_c^W(B^+, B_3)$	0.9232

TABLE 6.10
Values of Weighted Similarity Measures $S^W(B^-, B_j)$

$S_E^W(B^-, B_1)$	0.9445	$S_e^W(B^-, B_1)$	0.9165	$S_c^W(B^-, B_1)$	0.8992
$S_E^W(B^-, B_2)$	0.9617	$S_e^W(B^-, B_2)$	0.9434	$S_c^W(B^-, B_2)$	0.9327
$S_E^W(B^-, B_3)$	0.9768	$S_e^W(B^-, B_3)$	0.9639	$S_c^W(B^-, B_3)$	0.9550

TABLE 6.11
Numerical Values of $S^w(A_j)$

$S_E^W(B_1)$	0.5107	$S_e^W(B_1)$	0.5163	$S_c^W(B_1)$	0.5196
$S_E^W(B_2)$	0.5001	$S_e^W(B_2)$	0.4997	$S_c^W(B_1)$	0.4993
$S_E^W(B_3)$	0.4948	$S_e^W(B_3)$	0.4926	$S_c^W(B_1)$	0.4915

techniques in software engineering is in the development of time and cost-saving automatic systems and processes. The development in the field of computational intelligence aims to improve the quality of software. The development in the knowledge of FSs and fuzzy systems is relevant for software engineers, especially for software quality prediction, software test cases and predictive maintenance. The proposed measures can be efficiently applied to software engineering problems in situations where the entire information is not available. However, it may be noted that such measures can handle linguistic information in fuzzy and intuitionistic fuzzy environments only.

6.7 CONCLUSION

The existing measures of similarity between IFSs have proven to be very useful for solving practical problems in the situations of fuzziness. At the same time, there is a scope to develop more efficient similarity measures with applications in the field of artificial intelligence. The current study introduces several exponential similarity measures for IFSs with applications in the field of computational intelligence techniques. It also provides the proof of validity of the proposed measures. In the proposed process, we define the weighted exponential IF similarity measures. Moreover, we introduce a measure for solving the problems of strategic business decision-making employing the weighted exponential IF similarity measures. Finally, two quantitative illustrations exemplify the application of artificial intelligence in solving the problem of selecting investment opportunities in a cost-effective manner.

6.8 FUTURE RESEARCH DIRECTIONS

In future research, the measures offered here can be extended for interval-valued intuitionistic fuzzy (IVIF) statistics. We will explore some applications of artificial intelligence in IVIF environment in our forthcoming study. It is understood that fuzzy logic provides a basis for intelligent behavior in real-life problems. These approaches can be applied to get optimal solutions for practical problems such as defense, corporate sector, public policies and manufacturing in a cost-effective manner (Bouchon-Meunier et al. 2002; Tomar and Ohlan 2014; Kaur et al. 2018). The new software systems designed on the basis of these logics can efficiently handle different engineering tasks.

REFERENCES

Abdullah, L. 2013. Fuzzy multi criteria decision making and its applications: A brief review of category. *Procedia-Social and Behavioral Sciences* 97: 131–36.

Agrawal, A.P., Choudhary, A., Kaur, A. and Pandey, H.M. 2019. Fault coverage-based test suite optimization method for regression testing: Learning from mistakes-based approach. *Neural Computing and Applications* 1–16. https://link.springer.com/content/pdf/10.1007/s00521-019-04098-9.pdf

Atanasov, K.T. 1986. Intuitionistic fuzzy sets *Fuzzy Sets and Systems* 20(1): 87–96.

Bien, Z., Bang, W.C., Kim, D.Y. and Han, J.S. 2002. Machine intelligence quotient: Its measurements and applications. *Fuzzy Sets and Systems* 127(1): 3–16.

Boran, F.E. and Akay, D. 2014. A biparametric similarity measure on intuitionistic fuzzy sets with applications to pattern recognition. *Information Sciences* 255: 45–57.

Bouchon-Meunier, B., Gutiérrez-Ríos, J., Magdalena, L. and Yager, R.R. eds. 2002. *Technologies for Constructing Intelligent Systems 1: Tasks, in: Studies in Fuzziness and Soft Computing* (Vol. 89). Springer; Physica-Verlag Heidelberg.

Bustince, H., and Burillo, P. 1996. Vague sets are intuitionistic fuzzy sets. *Fuzzy Sets and Systems* 79(3): 403–05.

Chen, S.M. 1995. Measures of similarity between vague sets. *Fuzzy Sets and Systems* 74(2): 217–23.

Chen, S.M. 1997. Similarity measures between vague sets and between elements. *IEEE Transactions on Systems, Man, and Cybernetics, Part B (Cybernetics)* 27(1): 153–58.

Chen, S.M. and Chang, C.H. 2015. A novel similarity measure between Atanassov's intuitionistic fuzzy sets based on transformation techniques with applications to pattern recognition. *Information Sciences* 291: 96–114.

Chen, S.M., Cheng, S.H. and Lan, T.C. 2016. Multicriteria decision making based on the TOPSIS method and similarity measures between intuitionistic fuzzy values. *Information Sciences* 367: 279–95.

Dengfeng, L. and Chuntian, C. 2002. New similarity measures of intuitionistic fuzzy sets and application to pattern recognitions. *Pattern Recognition Letters* 23(1–3): 221–25.

Dubois, D. and Prade, H. 1991. Fuzzy sets in approximate reasoning, Part 1: Inference with possibility distributions. *Fuzzy Sets and Systems* 40(1): 143–202.

Farhadinia, B. 2014. An efficient similarity measure for intuitionistic fuzzy sets. *Soft Computing* 18(1): 85–94.

Garg, H. and Kumar, K. 2018. An advanced study on the similarity measures of intuitionistic fuzzy sets based on the set pair analysis theory and their application in decision making. *Soft Computing* 22(15): 4959–70.

Gau, W.L. and Buehrer, D.J., 1993. Vague sets. *IEEE Transactions on Systems, Man, and Cybernetics* 23(2): 610–14.

Hong, D.H. and Kim, C. 1999. A note on similarity measures between vague sets and between elements. *Information Sciences* 115(1–4): 83–96.

Horiuchi, K., Šešelja, B. and Tepavčević, A. 2020. Trice-valued fuzzy sets: Mathematical model for three-way decisions. *Information Sciences* 507: 574–84.

Hung, W.L. and Yang, M.S. 2004. Similarity measures of intuitionistic fuzzy sets based on Hausdorff distance. *Pattern Recognition Letters* 25(14): 1603–11.

Hung, W.L. and Yang, M.S. 2007. Similarity measures of intuitionistic fuzzy sets based on Lp metric. *International Journal of Approximate Reasoning* 46(1): 120–36.

Hung, W.L. and Yang, M.S. 2008a. On similarity measures between intuitionistic fuzzy sets. *International Journal of Intelligent Systems* 23(3): 364–83.

Hung, W.L. and Yang, M.S. 2008b. On the J-divergence of intuitionistic fuzzy sets with its application to pattern recognition. *Information Sciences* 178(6): 1641–50.

Hwang, C.M. and Yang, M.S. 2012. Modified cosine similarity measure between intuitionistic fuzzy sets. In *International Conference on Artificial Intelligence and Computational Intelligence*, 285–93. Berlin, Heidelberg: Springer.

Julian, P., Hung, K.C. and Lin, S.J. 2012. On the Mitchell similarity measure and its application to pattern recognition. *Pattern Recognition Letters* 33(9):1219–23.

Kaur, I., Narula, G.S., Wason, R., Jain, V. and Baliyan, A. 2018. Neuro fuzzy—COCOMO II model for software cost estimation. *International Journal of Information Technology* 10(2): 181–87.

Li, D.F. 2005. Multiattribute decision making models and methods using intuitionistic fuzzy sets. *Journal of Computer and System Sciences* 70(1): 73–85.

Li, F. and Xu, Z.Y. 2001. Measures of similarity between vague sets. Journal of Software 12(6): 922–927 (in Chinese).

Li, J., Deng, G., Li, H. and Zeng, W. 2012. The relationship between similarity measure and entropy of intuitionistic fuzzy sets. *Information Sciences* 188: 314–21.

Li, D.F. 2014. *Decision and Game Theory in Management with Intuitionistic Fuzzy Sets*, 308, 1–441. Berlin: Springer.

Liang, Z. and Shi, P. 2003. Similarity measures on intuitionistic fuzzy sets. *Pattern Recognition Letters* 24(15): 2687–93.

Liu, H.W. 2005. New similarity measures between intuitionistic fuzzy sets and between elements. *Mathematical and Computer Modelling* 42(1–2): 61–70.

Malhotra, R. and Bansal, A.J. 2015. Fault prediction considering threshold effects of object-oriented metrics. *Expert Systems* 32(2): 203–19.

Mishra, A.R., Jain, D. and Hooda, D.S. 2017. Exponential intuitionistic fuzzy information measure with assessment of service quality. *International Journal of Fuzzy Systems* 19(3): 788–98.

Mitchell, H.B. 2003. On the Dengfeng–Chuntian similarity measure and its application to pattern recognition. *Pattern Recognition Letters* 24(16): 3101–4.

Ohlan, A. 2015. A new generalized fuzzy divergence measure and applications. *Fuzzy Information and Engineering* 7(4): 507–23.

Ohlan, A. 2016. Intuitionistic fuzzy exponential divergence: application in multi-attribute decision making. *Journal of Intelligent & Fuzzy Systems* 30(3): 1519–30.

Ohlan, A. and Ohlan, R. 2016. *Generalizations of Fuzzy Information Measures.* Springer, Berlin, Heidelberg.

Park, J.H., Park, J.S., Kwun, Y.C. and Lim, K.M. 2007. New similarity measures on intuitionistic fuzzy sets. In *Fuzzy Information and Engineering*, 22–30. Springer, Berlin, Heidelberg.

Song, Y., Wang, X., Lei, L. and Xue, A. 2015. A novel similarity measure on intuitionistic fuzzy sets with its applications. *Applied Intelligence* 42(2): 252–61.

Szmidt, E. 2014. *Distances and Similarities in Intuitionistic Fuzzy Sets.* Springer International Publishing, Berlin, Heidelberg.

Tan, C. and Chen, X. 2014. Dynamic Similarity Measures between Intuitionistic Fuzzy Sets and Its Application. *International Journal of Fuzzy Systems* 16(4): 511–19.

Tomar, V.P. and Ohlan, A. 2014. New parametric generalized exponential fuzzy divergence measure. *Journal of Uncertainty Analysis and Applications* 2(24): 1–14.

Vlachos, I.K. and Sergiadis, G.D. 2007. Intuitionistic fuzzy information–applications to pattern recognition. *Pattern Recognition Letters*, 28(2): 197–206.

Wang, W. and Xin, X. 2005. Distance measure between intuitionistic fuzzy sets. *Pattern Recognition Letters* 26(13): 2063–69.

Wei, G.W. 2008. Maximizing deviation method for multiple attribute decision making in intuitionistic fuzzy setting. *Knowledge-Based Systems* 21(8): 833–836.

Wei, C.P., Wang, P. and Zhang, Y.Z. 2011. Entropy, similarity measure of interval-valued intuitionistic fuzzy sets and their applications. *Information Sciences* 181(19): 4273–86.

Wei, G. and Wei, Y. 2018. Similarity measures of Pythagorean fuzzy sets based on the cosine function and their applications. *International Journal of Intelligent Systems* 33(3): 634–52.

Xu, Z. 2007. Some similarity measures of intuitionistic fuzzy sets and their applications to multiple attribute decision making. *Fuzzy Optimization and Decision Making* 6(2): 109–21.

Yao, Y. 2020. Three-way granular computing, rough sets, and formal concept analysis. *International Journal of Approximate Reasoning*, 116: 106–25.

Ye, J. 2011. Cosine similarity measures for intuitionistic fuzzy sets and their applications. *Mathematical and Computer Modelling* 53(1–2): 91–97.
Zadeh, L.A. 1965. Information and control. *Fuzzy Sets* 8(3): 338–53.
Zadeh, L.A. 1968. Probability measures of fuzzy events. *Journal of Mathematical Analysis and Applications* 23(2): 421–27.
Zadeh, L.A. 2008. Is there a need for fuzzy logic? *Information Sciences* 178(13): 2751–79.
Zhang, C. and Fu, H. 2006. Similarity measures on three kinds of fuzzy sets. *Pattern Recognition Letters* 27(12): 1307–17.

7 Nature-Inspired Approaches to Test Suite Minimization for Regression Testing

Anu Bajaj and Om Prakash Sangwan
Guru Jambheshwar University of Science and Technology

CONTENTS

7.1 INTRODUCTION

Nowadays, software industries use continuous integration and continuous development models, with new requirements being added along with their updates. Regression testing plays a crucial role in retaining quality and reliability (Yamuç, Cingiz, Biricik, and Kalıpsız 2017). Regression testing is performed to check whether the changes introduced any faults or vulnerabilities. The extensive collection of test cases takes more time ranging from days to weeks to run (Zheng, Hierons, Li, Liu, and Vinciotti 2016). In addition, new and updated requirements need more test cases to be designed for complete testing, which creates a broad set of redundant test cases (Zhang, Liu, Cui, Hei, and Zhang 2011). Consequently, the regression test suite reduction came into practice. This approach tries to remove redundancy, which thereby reduces the cost of execution. The competitive market strives for small budgets and quick solutions (Yamuç, Cingiz, Biricik, and Kalıpsız 2017), which require a representative set as efficient as the original set. Computational intelligence algorithms are capable of solving such complex problems. This problem is equivalent to a minimal set cover problem, that is, NP-complete problem. Nature-inspired algorithms have the potential to solve NP-complete problems efficiently and effectively (Haider, Nadeem, and Akram 2016). In this chapter, we focus on test suite minimization approaches that use nature-inspired algorithms. We formulated the following research questions to assess the potential of nature-inspired algorithms in solving the problem:

RQ1: Which nature-inspired optimization algorithms are used in the field of regression test suite minimization?
RQ2: Whether the optimization algorithms are used for a single objective or multi-objective test suite minimization?
RQ3: Which testing criteria are used for designing fitness function?
RQ4: Which performance metrics and datasets are used for validation of the results?

The chapter is organized as follows: first, we briefly introduce the test suite minimization problem and nature-inspired algorithms. Second, the research work performed in this field is described in detail. Third, we address the research questions and future directions with the help of the observations and findings of the related studies. Finally, the chapter findings have been summarized in conclusion.

7.2 REGRESSION TEST SUITE MINIMIZATION

Test suite minimization finds subsets of test suites to remove redundant and obsolete test cases. Yoo and Harman (2010) described the problem as:

Given: A test suite T of n test cases $T_1, T_2,..., T_n$, covering m requirements $R=R_1, R_2,..., R_m$.

Problem: To find a subset T' such that $T'<T$ and T' is capable of satisfying all the requirements of set R.

These test case requirements could be any test case requirements, for example, code coverage, du pairs and condition coverage. The test requirement can be satisfied by any test case, but it should be satisfied by at least one test case. Figure 7.1 shows an example of test cases and requirements covered by these test cases. The subsets could be {$T3$, $T6$}, {$T1$, $T4$}, {$T4$, $T6$}, and so on. Therefore, the representative set problem is the minimal hitting problem, which is best solved by computational algorithms (Yoo and Harman 2010).

7.3 NATURE-INSPIRED ALGORITHMS

Computational intelligence algorithms are becoming more popular among modern numerical optimization methods. Real-world problems are becoming more and more

Test Cases	Requirements				
	R1	R2	R3	R4	R5
T1	1	0	0	1	1
T2	0	0	0	1	1
T3	0	0	1	0	0
T4	0	1	1	0	0
T5	0	1	0	1	1
T6	1	1	0	1	1

FIGURE 7.1 Test cases versus requirements example.

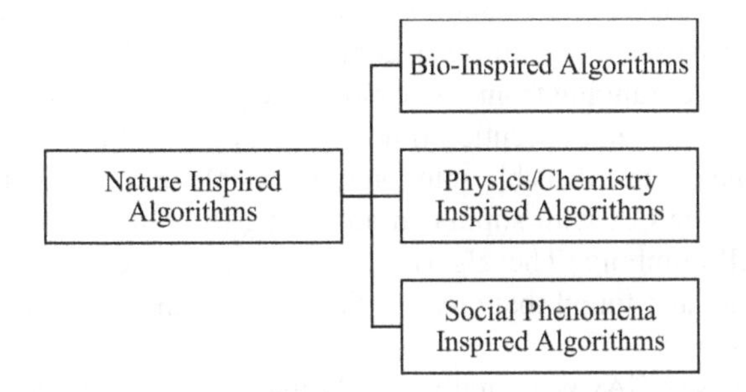

FIGURE 7.2 Nature-inspired algorithms.

complex, that is, NP-complete problems. Researchers have invested significant effort in solving these problems with the help of natural phenomena. There are various sources of inspiration, such as biology, physics, chemistry and social phenomena, as shown in Figure 7.2 (Bajaj and Sangwan 2014). The algorithms that mathematically formulate the problem by mimicking the natural phenomena and provide the optimal solutions are known as nature-inspired algorithms (Fister, Yang, Fister, Brest, and Fister 2013). Most of the studies have used biology-inspired algorithms compared to physics- and chemistry-inspired algorithms. Algorithms such as genetic algorithm (GA) and particle swarm optimization (PSO) are well-known algorithms. They have also proven their efficiency in solving regression testing problems, such as broader applicability of GAs in prioritizing test cases (Bajaj and Sangwan 2019). The next section explains nature-inspired algorithms used in reducing the test suite size.

7.4 NATURE-INSPIRED ALGORITHMS USED IN REGRESSION TEST SUITE MINIMIZATION

Various researchers have used nature-inspired algorithms for solving the regression test suite minimization. A brief description of these works is presented below.

7.4.1 Genetic Algorithms (GA)

Ma, Sheng, and Ye (2005) provided the mathematical model for mapping the test suite minimization problem to optimization algorithms. They applied GA on two small programs, Calculator and Little TV. The GA outperformed the modified greedy algorithm to suite cost, suite size and execution time of the algorithms. Zhong, Zhang, and Mei (2006) empirically evaluated the performance of GA with traditional heuristics H, GRE and ILP on small-to-large size programs. They observed that GA is weaker than these conventional algorithms in terms of suite size and execution time. Ma, Zhao, Liang, and Yun (2010) also empirically analyzed these algorithms on large suites and ranked these algorithms by varying the requirements and suite size. They also found that the GA performed worse than the other five algorithms. This is possibly due to its tendency to stick to local optima.

Several authors modified the GA to enhance its efficiency and effectiveness for test suite reduction. Zhang, Liu, Cui, Hei and Zhang (2011) applied adaptive quantum-inspired GA using quantum encoding and cross-referencing crossover operator. The proposed algorithm was compared with H, GRE, GA and quantum GA for the test cases ranging from 10 to 5000. The proposed algorithm gave a better reduction in suite size and execution time. You and Lu (2012) used a repair operator in GA to convert an infeasible solution to a feasible one. Mansour and El-Fakih 1999 hybridized the GA with simulated annealing (SA) by introducing feasibility, penalty and hill-climbing. The algorithms were compared with the incremental approach, slicing, and found that GA and SA miss useful tests and have lesser tendency comparatively.

The weight-based GAs were applied by Wang, Ali, and Gotlieb (2013) for software product line testing. The objectives were test minimization percentage (TMP), fault detection capability (FDC) and feature pairwise coverage (FPC). The experiment performed on Cisco and SPLOT projects showed that random-weight GA (RWGA) was better than weight-based GA (WBGA). They also claimed that the performance of the optimization algorithm is affected by the inappropriate fitness function design and different parameter settings. Furthermore, Wang, Ali, and Gotlieb (2015) experimented with 10 different nature-inspired algorithms, including CellDE, WBGA, PAES, RWGA, SMPSO, SPEA2 and MoCell. They observed that RWGA outperformed all algorithms while considering all the objectives together.

Pedemonte, Luna, and Alba (2016) implemented GA with parallelism, that is, systolic GA for test suite minimization of very large test suites. They also introduced the penalty fitness function by penalizing the uncovered requirements of the current solution. The speedup and correlation values were higher for systolic GA than simple GA and elitist GA. To increase the speed of the computation for huge test suites, Yoo, Harman, and Ur (2011) proposed highly scalable multi-objective minimization using GPGPU. Furthermore, they experimented with SPEA2, Two Archive EA (TAEA) (Yoo, Harman, and Ur 2013). They observed that all algorithms provided speedup of about 25 times, and that the performance of the optimization algorithm increased with the increase in the size of the program and test suite.

Yoo and Harman (2010) performed the first experiment on multi-objective test suite minimization. They used a hybrid of additional greedy and NSGA-II and formulated two and three objective problems considering fault coverage, statement coverage and execution time. They observed that additional greedy performed better than the multi-objective NSGA-II, and found a correlation between the past fault coverage and code coverage.

Marchetto, Scanniello, and Susi (2017) proposed modified MOEA and used cumulative code coverage, requirement coverage and relative cost as objective functions. The experiment performed on 20 subject programs and the performance comparison was based on various metrics, such as size reduction, time reduction and diversity. The proposed algorithm was cost-effective but not efficient in reducing suite size compared to seven baseline approaches. Zheng, Hierons, Li, Liu, and Vinciotti (2016) aimed at another MOEA with displacement (MOEA/D) with fixed and varying c value. It was observed that variable c MOEA/D performed better than others. NSGA-II was used for GUI software test suite reduction by considering two objectives, event coverage and fault coverage (Chaudhary and Sangwan 2016). Geng, Li, Zhao, and Guo (2016) incorporated fault localization along with fault detection to localize and simultaneously detect the faults.

Several studies explored nature-inspired algorithms other than GA for test suite minimization, for example, ant colony optimization (ACO), BAT algorithm, Dragonfly algorithm (DA), differential evolution (DE) and neural networks.

7.4.2 Differential Evolution (DE)

Kumari, Srinivas, and Gupta (2012) used a quantum-inspired DE algorithm for multi-objective test suite reduction considering the cost of coverage and execution. They evaluated the Pareto fronts size, the number of solutions obtained and the run time of the algorithm in real-world programs. The proposed algorithm produced high-quality solutions and more number of solutions compared to NSGA-II.

7.4.3 Ant Colony Optimization (ACO)

Mohapatra and Prasad (2015) implemented the ACO algorithm for test suite reduction. They experimented on five Java Programs and compared the proposed algorithm with the traditional methods, Harold's Heuristic, GRE Heuristic, GA and ILP models, for suite size and complexity of test suite. It was observed that ACO performed better than other algorithms for efficiency and effectiveness. Zhang, Yang, Lin, Dai, and Li (2017) also proposed a test suite minimization method using ACO. However, they used a quantum encoding scheme for test case representation to maximize requirement coverage and minimize cost. They compared the performance with greedy algorithms and other algorithms of the same family, that is, ACO, ant system and maximal-minimal ant colony system. The quantum-inspired ACO outperformed the compared algorithms.

7.4.4 BAT Algorithm

Sugave, Patil, and Reddy (2018) applied diversity-based BAT algorithm. They injected diversity within the solutions and between the solutions using the bitwise OR operator. The authors also proposed a new measure, adaptive test case pool measure (ATAP), and defined a new objective function. The validation of the algorithm done with the help of nine evaluation metrics like suite cost, suite size and common rate, etc., as well as eight software programs of SIR repository. The algorithm performed better compared to BAT, greedy, diversity-based GA, TAP and ATAP.

7.4.5 Dragonfly Algorithm

Sugave, Patil, and Reddy (2017) also proposed a diversity-based DA. They compared the effectiveness for suite cost, suite size, improvement in suite cost and suite size to greedy, systolic GA and diversity BAT. The results showed DA outperformed other algorithms.

7.4.6 Neural Network

Test suite reduction with optimization algorithms leads to a dramatic decrease in suite size, which may miss useful test cases and, subsequently, fault detection rate. Anwar, Ahsan, and Catal (2016) tried to overcome this limitation by incorporating human judgment with the help of neuro-fuzzy modeling. They considered the fault detection rate, requirement failure impact, execution time and requirement coverage. This system suffers from parameter tuning, which was solved by Anwar, Afzal, Bibi, Abbas, Mohsin, and Arif (2019) by hybridizing the adaptive neuro-fuzzy inferencing system (ANFIS) with GA and PSO. They compared them with NSGA-II, ANFIS and TOPSIS and found that the results were better comparatively. Haider, Nadeem, and Akram (2016) also implemented ANFIS and showed that the system could not replace human judgment optimality. However, these systems are helpful for larger test suites. Table 7.1 outlines the nature-inspired approaches to test suite minimization for regression testing.

7.5 FINDINGS AND FUTURE DIRECTIONS

The research questions are answered below in the form of findings and future directions.

RQ1: Nature-inspired algorithms used in test suite minimization

Findings: GAs are the most widely used algorithms for the test suite minimization problem, possibly because of its popularity and easy access. Some authors have claimed that GA does not perform as well as traditional algorithms. However, other researchers have shown better performance and have used multi-objective evolutionary algorithms and their parallel versions to reduce the test suite size and cost. This is possibly because either the experiment was performed on different subject programs or the authors used different objective functions for evaluation.

Future Directions: Very few studies have used other nature-inspired algorithms, such as ACO, BAT, DA, DE and PSO. Hybrid algorithms can offer more efficient solutions, as is evident from the performance of hybrid ANFIS, hybrid NSGA-II and hybrid GASA. Relatively new algorithms or other categories of nature-inspired algorithms such as physics-/chemistry-inspired optimization algorithms are yet to be explored for regression test suite minimization. In other words, there is still a potential gap that can be filled in the future.

TABLE 7.1
Test Suite Minimization Using Nature-Inspired Algorithms

Author Year	Nature-Inspired Algorithms Used	Single or Multi-Objective	Optimization Measure	Performance Metrics	Datasets	Compared With
Mohapatra and Prasad (2015)	ACO	Single	Maximize code coverage	Suite size reduction and complexity of test suite	Binary search tree, stack, power equalizer, transmission control, stock	Harold's heuristic, GRE heuristic, GA, ILP models
Zhang, Yang, Lin, Dai and Li (2017)	Quantum ACO	Single	Maximize requirement coverage and minimize cost	Execution time	SIR repository programs	ACO, ant colony system, greedy, maximal-minimal ant System
Sugave, Patil, and Reddy (2018)	Diversity BAT algorithm	Single	Full coverage with minimum cost	Cost reduction and improvement in cost percentage, average improvement percentage on cost, suite size, and iterations	SIR repository programs	Greedy EIrreplacability, systolic genetic algorithm (GA), test pool measure (TAP), ATAP
Sugave, Patil and Reddy (2017)	Diversity Dragonfly algorithm	Single	Full coverage with minimum cost	Cost reduction and improvement in cost percentage	SIR repository programs	Greedy EIrreplacability, systolic genetic algorithm (GA)
Kumari, Srinivas and Gupta (2012)	Multi-objective Quantum-inspired DE	Multi-objective	Code coverage and execution cost	Pareto fronts size, number of solutions, and average execution time of algorithms	Two real case studies	NSGA-II
Ma, Sheng, and Ye (2005)	GA	Single	Code coverage	Suite size reduction, cost reduction, and execution time	Calculator and little TV	Modified greedy algorithm

(Continued)

TABLE 7.1 (*Continued*)
Test Suite Minimization Using Nature-Inspired Algorithms

Author Year	Nature-Inspired Algorithms Used	Single or Multi-Objective	Optimization Measure	Performance Metrics	Datasets	Compared With
Zhong, Zhang, and Mei (2006)	GA	Single	Code coverage	Suite size reduction, execution time	SIR repository programs	Harold's heuristic, GRE heuristic, and ILP model
Ma, Zhao, Liang, and Yun (2010)	GA	Single	Maximize requirement coverage	Suite size reduction	Private programs	Greedy, Harold's heuristic, GRE heuristic, GE, and ILP model
Mansour and El-Fakih (1999)	Hybrid GA SA	Single	Maximize modified statement coverage	Modified statement coverage and execution time	Private Programs	Incremental approach, slicing, SA
Wang, Ali, and Gotlieb (2013)	Weight-based GAs	Multi-objective	Maximize FPC, FDC, and minimize TMP	FPC, FDC, and TMP	Industrial and SPLOT case studies	Random search
Pedemonte, Luna and Alba (2016)	Systolic GA	Multi-objective	Code coverage and execution cost	Cost reduction and increase in speed up	SIR repository programs	GA, elitist GA
Anwar, Afzal, Bibi, Abbas, Mohsin and Arif (2019)	ANFIS-GA, ANFIS-PSO	Multi-objective	Maximize requirement coverage, fault detection rate (FDR), and minimize execution time, requirement failure impact (RFI)	Percentage loss in FDR, RC, Percentage reduction in suite size and cost	SIR repository program and previous data problem	TOPSIS, GA, NSGA-II, MOPSO

(*Continued*)

TABLE 7.1 (*Continued*)
Test Suite Minimization Using Nature-Inspired Algorithms

Author Year	Nature-Inspired Algorithms Used	Single or Multi-Objective	Optimization Measure	Performance Metrics	Datasets	Compared With
Yoo, Harman and Ur (2010)	Parallel NSGA-II	Multi-Objective	Maximize statement coverage and minimize time	Speed-up and correlation	SIR repository programs	CPU and GPU implementation of NSGA-II
Zheng, Hierons, Li, Liu and Vinciotti (2016)	MOEA/D	Multi-objective	Maximize statement, branch, conditions coverage, and minimize cost	Hypervolume	SIR repository programs	Greedy, NSGA-II
Marchetto, Scanniello and Susi (2017)	NSGA-II	Multi-objective	Maximize statement and requirement coverage, and minimize cost	Suite size reduction, cost reduction, execution time, diversity, co-factor analysis percentage loss in FDR	Java programs	Greedy, delayed greedy, 2 opt greedy, Harrold Gutpa Soffa Heuristic, 3D greedy

RQ2. Single versus multi-objective optimization

Findings: Most studies used a single objective optimization problem with the main focus either on maximizing coverage or minimizing cost. Few studies implemented multi-objective optimizations, and they also experimented with evolutionary algorithms.

Future Directions: There is a need to conduct more studies on multi-objective optimization by considering two or more objectives simultaneously as, in real-world, we have many goals and constraints. Therefore, multi-objective optimization is another area of research for solving the test suite minimization problem.

RQ3. Fitness function design and parameter settings

Findings: Maximizing coverage and minimizing cost are the testing criteria used by most researchers for solving the problem. We also observed that a few studies have empirically analyzed the effect of parameter settings on the efficiency and effectiveness of algorithms.

Future Directions: Designing a suitable fitness function and optimizing the parameter settings of algorithms is challenging. We need to shift the focus to other criteria like requirement priorities, changes, failure rates and adding more cost-effectiveness measures. More studies are required to validate the combination of testing criteria and analyze parameter settings.

RQ4. Performance Metrics and Datasets

Findings: The commonly used evaluation metrics were percentage suite size reduction, percentage cost reduction and percentage loss in fault detection. Hypervolume, Pareto fronts size and the number of solutions obtained for the multi-objective optimization problem. The experiments were either performed on SIR repository programs or private programs.

Future Directions: Most researchers performed experimental studies. A few studies have used real-world industrial applications. There is a need to conduct empirical studies on large and more real-world applications to generalize the results.

7.6 CONCLUSION

The software keeps on evolving day by day, and hence the vast collection of retests is challenging to execute exhaustively. Elimination of redundant and obsolete test cases without affecting the effectiveness leads to a minimal set cover problem, that is, NP-complete problem. These types of issues can be efficiently solved with the help of optimization algorithms. In this chapter, we identified studies whose focus is on the use of nature-inspired algorithms for test suite minimization. We found that the most popular algorithms are GA, and NSGA-II and relatively new and hybrid algorithms are not yet explored. Widely used fitness functions are maximizing code coverage and minimizing execution costs. Most studies used single objective optimization and default parameter settings. In conclusion, nature-inspired algorithms are capable of solving the regression test minimization problems, and the field has scope for further research on multi-objective optimization and parameter tuning. In the future, we will experiment on relatively new or hybrid algorithms for regression test suite minimization.

ACKNOWLEDGMENT

This work was supported by the University Grants Commission (UGC) of the Government of India under the Senior Research Fellowship (SRF) Scheme with reference number 3469/ (NET-DEC. 2014).

REFERENCES

Anwar, Z., Afzal, H., Bibi, N., Abbas, H., Mohsin, A., and Arif, O. 2019. A hybrid-adaptive neuro-fuzzy inference system for multi-objective regression test suites optimization. *Neural Computing and Applications* 31, no. 11: 7287–7301.

Anwar, Z., Ahsan, A., and Catal, C. 2016. Neuro-fuzzy modeling for multi-objective test suite optimization. *Journal of Intelligent Systems* 25, no. 2: 123–146.

Bajaj, A., & Sangwan, O. P. 2019. A systematic literature review of test case prioritization using genetic algorithms. *IEEE Access* 7: 126355–126375.

Bajaj, A., and Sangwan, O. P. 2018. A survey on regression testing using nature-inspired approaches. In *2018 4th International Conference on Computing Communication and Automation:* 1–5. IEEE.

Chaudhary, N., and Sangwan, O. P. 2016. Multi Objective Test Suite Reduction for GUI Based Software Using NSGA-II. *International Journal of Information Technology and Computer Science 8*: 59–65.

Fister J, I., Yang, X.S., Fister, I., Brest, J. and Fister, D. 2013. A brief review of nature-inspired algorithms for optimization *arXiv preprint arXiv*:1307no. 4186: 116–122.

Geng, J., Li, Z., Zhao, R., and Guo, J. 2016. Search based test suite minimization for fault detection and localization: A co-driven method. In *International Symposium on Search Based Software Engineering*, 34–48. Springer, Cham.

Haider, A. A., Nadeem, A., and Akram, S. 2016. Regression test suite optimization using adaptive neuro fuzzy inference system. In *2016 International Conference on Frontiers of Information Technology,* 52–56. IEEE.

Kumari, A. C., Srinivas, K., and Gupta, M. P. 2012. Multi-objective test suite minimisation using quantum-inspired multi-objective differential evolution algorithm. In *2012 IEEE International Conference on Computational Intelligence and Computing Research*, 1–7. IEEE.

Ma, X. Y., Sheng, B. K., and Ye, C. Q. 2005. Test-suite reduction using genetic algorithm. In *International Workshop on Advanced Parallel Processing Technologies*, 253–262. Springer, Berlin, Heidelberg.

Ma, Y., Zhao, Z., Liang, Y., and Yun, M. 2010. A usable selection range standard based on test suite reduction algorithms. *Wuhan University Journal of Natural Sciences* 15, no. 3: 261–266.

Mansour, N., and El-Fakih, K. 1999. Simulated annealing and genetic algorithms for optimal regression testing. *Journal of Software Maintenance: Research and Practice*, 11, no. 1: 19–34.

Marchetto, A., Scanniello, G., and Susi, A. 2017. Combining code and requirements coverage with execution cost for test suite reduction. *IEEE Transactions on Software Engineering* 45, no. 4: 363–390.

Mohapatra, S. K., and Prasad, S. 2015. Test Case Reduction Using Ant Colony Optimization for Object Oriented Program. *International Journal of Electrical & Computer Engineering* 5, no. 6: 2088–8708.

Pedemonte, M., Luna, F., and Alba, E. 2016. A Systolic Genetic Search for reducing the execution cost of regression testing. *Applied Soft Computing*, 49: 1145–1161.

Sugave, S. R., Patil, S. H., and Reddy, B. E. 2017. DDF: Diversity dragonfly algorithm for cost-aware test suite minimization approach for software testing. In *2017 International Conference on Intelligent Computing and Control Systems,* 701–707. IEEE.

Sugave, S. R., Patil, S. H., and Reddy, B. E. 2018. DIV-TBAT algorithm for test suite reduction in software testing. *IET Software* 12, no. 3: 271–279.

Wang, S., Ali, S., and Gotlieb, A. 2013. Minimizing test suites in software product lines using weight-based genetic algorithms. In *Proceedings of the 15th Annual Conference on Genetic and Evolutionary Computation*, 1493–1500.

Wang, S., Ali, S., and Gotlieb, A. 2015. Cost-effective test suite minimization in product lines using search techniques. *Journal of Systems and Software*, 103: 370–391.

Yamuç, A., Cingiz, M. Ö., Biricik, G., and Kalıpsız, O. 2017. Solving test suite reduction problem using greedy and genetic algorithms. In *2017 9th International Conference on Electronics, Computers and Artificial Intelligence*, 1–5. IEEE.

Yoo, S., and Harman, M. 2010. Using hybrid algorithm for pareto efficient multi-objective test suite minimisation. *Journal of Systems and Software* 83, no. 4: 689–701.

Yoo, S., Harman, M., and Ur, S. 2011. Highly scalable multi objective test suite minimisation using graphics cards. In *International Symposium on Search Based Software Engineering*, 219–236. Springer, Berlin, Heidelberg.

Yoo, S., Harman, M., and Ur, S. 2013. GPGPU test suite minimisation: Search based software engineering performance improvement using graphics cards. *Empirical Software Engineering* 18, no. 3: 550–593.

You, L., and Lu, Y. 2012. A genetic algorithm for the time-aware regression testing reduction problem. In *2012 8th International Conference on Natural Computation*, 596–599. IEEE.

Zhang, Y. K., Liu, J. C., Cui, Y. A., Hei, X. H., and Zhang, M. H. 2011. An improved quantum genetic algorithm for test suite reduction. In *2011 IEEE International Conference on Computer Science and Automation Engineering* vol. 2, 149–153. IEEE.

Zhang, Y. N., Yang, H., Lin, Z. K., Dai, Q., and Li, Y. F. 2017. A test suite reduction method based on novel quantum ant colony algorithm. In *2017 4th International Conference on Information Science and Control Engineering,* 825–829. IEEE.

Zheng, W., Hierons, R. M., Li, M., Liu, X., and Vinciotti, V. 2016. Multi-objective optimisation for regression testing. *Information Sciences*, 334: 1–16.

Zhong, H., Zhang, L., and Mei, H. 2006. An experimental comparison of four test suite reduction techniques. In *Proceedings of the 28th International Conference on Software Engineering*, 636–640.

8 Identification and Construction of Reusable Components from Object-Oriented Legacy Systems Using Various Software Artifacts

Amit Rathee and Jitender Kumar Chhabra
National Institute of Technology

CONTENTS

8.1 INTRODUCTION

The research in the field of software engineering is mainly perturbed in view of the foregoing development approaches that aim at reducing development cost and time and improving the software's quality. One such emergent software development technique that focuses on the development of software by integrating independent and reusable components is termed as component-based software development (CBSD), which provides systematic reuse opportunities. The modern software industries are shifting to the CBSD paradigm for the rapid building of large and complex software systems with minimal efforts required at the developer's end (Kahtan et al. 2016). The concept of component reuse in CBSD is derived from the reusability already conducted in the engineering and manufacturing companies (Qureshi and Hussain 2008).

The component-based software engineering (CBSE) approach consists of three necessary and mandatory parallel engineering activities that are the key in deciding the overall success of CBSE in software development. These activities are depicted in Figure 8.1 and are termed as component library construction, domain analysis and component-based development (CBD). The component library construction activity is engaged with the creation of component repositories/libraries. The component library can be constituted in three ways, namely, components can be developed and stored from scratch, components can be bought from third parties in the form of COTS and components can be identified using high-quality source code available with legacy software. The domain analysis aspect of CBSE involves the analysis of requirements laid down for the new software system and identification of required components from component repositories. CBD involves the integration of identified components for obtaining new software systems.

The benefits of the CBSD approach are limited and widely dependent on the availability of quality components in the component repository (Rathee and Chhabra

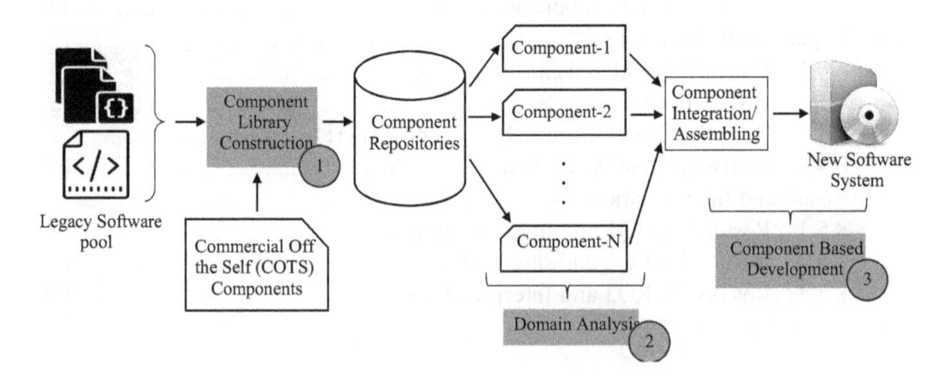

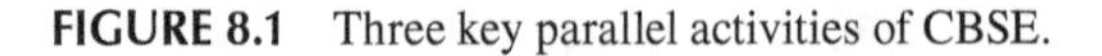
FIGURE 8.1 Three key parallel activities of CBSE.

2019b). Rathee and Chhabra (2019a) are of the opinion that such a high-quality repository can be constituted after identifying reusable components using legacy software system's source code. Such legacy software systems provide attractive reuse opportunities because of the availability of already tested, validated, verified functionalities (Hein 2014, Rathee and Chhabra 2019c). Moreover, using existing source code to build reusable components helps in reducing development cost involved with the cases where the components are either built from scratch or purchased from third-party vendors. This statement is backed by the fact that the U.S. Department of Defense alone can annually save a sum of $300 million by only enhancing the software reuse by a factor of 1% (Anthes 1993). Thus, identifying components from the legacy software system is a crucial, cost-effective and mandatory step for the successful implementation of the CBSD approach.

Many closed and relevant approaches are already proposed in the literature that identify components using an object-oriented source code of legacy systems (El Hamdouni et al. 2010, Shatnawi and Seriai 2013, Kebir et al. 2012, Allier et al. 2011, Alshara et al. 2015). These approaches group different software elements such that the group provides unique functionality, and this group is referred to as a component. However, there are several limitations to these techniques. First, identifying the mandatory part of the component (its provides and requires interface) is not targeted by some of the approaches (El Hamdouni et al. 2010). However, a component's interface identification is vital to explicitly utilize its services by other components. Second, many of these approaches mandate the selection of different client applications or the importance of semantic relations is forgotten during the identification of components. In almost all the above techniques, a component is described only informally in terms of a set of software elements that are a part of it along with the description of its corresponding interfaces (Yacoub et al. 1999). Specifying a component only informally does not make it directly reusable. To make them directly reusable, they must be transformed as per the recommendations of some component models, namely, JavaBeans, Microsoft COM, OMG's CORBA and EJB (Enterprise Java Bean) (Heisel et al. 2002). Such a transformed component can be directly integrated with other components and they can be directly introspected by third-party tools. However, it is the belief of the authors that no technique available in the literature targets modeling the component as JavaBeans.

Therefore, this chapter discusses two main important facets of the CBSE approach. The first considered key problem is the identification of high-quality components from the object-oriented software. This chapter gives due importance to the semantic integrity aspect of a component along with other structural relations. Here, clustering is modeled as search-based optimization using the NSGA-II algorithm. The second key challenge taken up by this chapter is the construction of a reusable physical component by the transformation of the identified logical component to the corresponding JavaBean (Abdellatief et al. 2013). The important research objectives of this chapter are outlined below:

1. The necessity and benefits for the IT industry of identifying reusable components from the source code of the legacy system.
2. The role of different software artifacts for the identification of components from legacy systems.

3. The role of search-based optimization techniques in the reusable component identification process.
4. The necessity of component construction through transformation for the logical components that are identified from the existing source code.

8.2 LITERATURE SURVEY

Since the past few decades, researchers and practitioners have actively targeted the CBSE and many researchers have proposed different approaches for component identification. This section provides a review of different techniques that are directly or indirectly related to component identification. Moreover, the strengths and weaknesses of these approaches are also presented.

Different researchers have proposed techniques that target software dependency relations estimation or utilization. The various possible types of structural relations which are generally available in an object-oriented software system are annotation, reference, call, parameters passed to a method, is of type, return, the number of method calls, has a class type parameter, implements, extends, throws and exceptions (Briand et al. 1999, Erdemir and Buzluca 2014, Maffort et al. 2016). According to Rathee and Chhabra (2018), three kinds of commonly used software dependencies in software remodularization are structural, conceptual and evolutionary dependency relations. Different researchers have applied the structural dependency analysis at different granularity levels of the software, namely, statements, methods, classes and architecture (Stafford and Wolf 2001, Vieira et al. 2001). Czibula et al. (2019) studied the maintenance and evolution problem in an existing software. They introduced a new aggregate coupling measurement scheme to ease the software maintenance problem by estimating the impact of changes. Rathee and Chhabra (2019c) identified reusability opportunities in the multimedia software system by utilizing various dependency relations among different software elements, namely, structural and semantic. They proposed a new structural dependency measurement scheme called frequent usage pattern (FUP).

Researchers have also actively targeted software remodularization. Software remodularization aids in reducing maintenance efforts, improving quality and ultimately reusing the software system (Ducasse and Pollet 2009). Bittencourt and Guerrero (2009) also conducted another empirical study to determine the efficiency of different clustering algorithms. They used the MoJoSim metric for the measurement of the authoritativeness of restructuring results. Candela (2016) empirically studied two quality attributes, namely, cohesion and coupling from the software remodularization viewpoint on 100 open-source systems. Based on the experimentation, the authors concluded that cohesion and coupling are not only parameters for an effective remodularization solution, refactoring efforts should be considered during remodularization, and existing tools that support "Big Bang" remodularization had limited practical applicability. Beck et al. proposed a visualization approach for the current modularization of a software system using different hierarchical clustering criteria (Beck et al. 2016). Corazza et al. suggested a technique that helps in dividing software into semantically related classes by using a hierarchical agglomerative clustering algorithm (Corazza et al. 2011). Chhabra (2017) proposed an approach to improve the modular structure by considering lexical and structural aspects and modeling the software remodularization problem as a single and multiobjective search-based problem. Abdeen et al. applied the multiobjective search-based approach for improving the existing software structure by performing software remodularization using NSGA-II (Abdeen et al. 2013). Barros et al. performed a case study to evaluate the applicability of search-based remodularization techniques in a software architecture recovery context with large-sized open-source software (de Oliveira Barros et al. 2015). They conclude that search-based module clustering algorithms are capable of recovering the degraded software architecture that is most similar to the design-time architecture of the software.

Software clustering is another research area that affects the reusable component identification process. Mohammadi and Izadkhah (2019) proposed a new deterministic software clustering algorithm called neighborhood tree algorithm for extracting meaningful and less complex subsystems. The proposed algorithm performs clustering using a neighborhood tree constructed from the artifact dependency graph. Chong and Lee (2017) proposed an automatic approach for identifying different clustering constraints using the graph theory concept for the software system. The identified constraints can help in achieving flexibility and together can minimize the supervision of the domain expert. Prajapati and Chhabra (2018) proposed a multi-objective two-archive-based artificial bee colony (TA-ABC) software module clustering algorithm. The proposed clustering algorithm uses improved indicator-based selection approach compared to the Pareto dominance selection technique.

Some researchers have also directly worked in the field of component identification. Szyperski et al. (2002) provided a definition of a software component that is widely being adopted among the research community. They stress the fact that the interfaces are the fundamental building blocks and each component may have properties, methods, events and relationships among the interfaces of the different components. Constantinou et al. (2015) proposed a semi-automated approach for identifying reusable software components by untangling complex structures of the software. They presented their approach in the form of a tool termed ReCompId. Shatnawi et al. (2018) reported an approach for converting object-oriented APIs into the corresponding component-based APIs based on dynamic relations. Chardigny et al. (2008) proposed the ROMANTIC approach for the component-based architecture recovery of a software system. The structural and semantic characteristics of the underlying software system are utilized by the ROMANTIC approach. Mishra et al. (2009) transformed an object-oriented system into its equivalent component-based system. Their approach is termed as component-oriented reverse engineering (CORE) model, and their approach is based on the clustering process and the CRUD matrix (Created, Deleted, Read, and Updated) patterns in a system.

Component modeling and transformation is another research field and is only targeted by few researchers. Lau et al. (2005) presented a taxonomy related to different software component models. Song et al. (2002) proposed a code analysis approach that converts identified components from the underlying Java Servlet source code to Enterprise JavaBeans components. Washizaki and Fukazawa (2005) proposed a refactoring technique for converting a Java software to its equivalent JavaBean system using the class relations-based dependency graph and Façade interfaces. Allier et al. (2011) reported a transformation technique for converting Java application to the OSGi component model standard using software design patterns, namely, Adapter and Façade.

Based on the literature review, very limited work is devoted to the component identification from legacy software systems along with transformation as per the recommendations of the component model. Further, no other studies except the one by Washizaki and Fukazawa (2005) targeted converting identified components to JavaBeans. The existing metaheuristic approaches are mainly focused on software remodularization that is different from the concept of reusable components. Moreover, there is a wide research gap in the field of dependency relations between different software elements of a legacy system that directly affect the quality of a component. Finally, to our knowledge, no existing work utilizes usage pattern, semantic and evolutionary relations together during reusable components identification from legacy systems.

8.3 PROPOSED COMPONENT IDENTIFICATION AND CONSTRUCTION APPROACH

Identifying reusable components from legacy software is an important and necessary step for the successful implementation and wide applicability of the CBSD approach. Therefore, this section discusses an approach that helps in identifying high-quality software components. Figure 8.2 shows the various steps involved in the reusable component identification from object-oriented legacy software systems. In the proposed approach, different elements of a legacy software system are initially considered independent and are then are iteratively clustered into independent groups that are finally termed as logical components. This step is shown as *Logical Components Identification* in Figure 8.2. Each identified logical component designates a unique functionality in the legacy system. Finally, these logical components are transformed into the corresponding physical components by identifying their interfaces and transforming them as per the recommendations of the JavaBeans component model (Englander 1997). This step is shown as *Component Construction* in Figure 8.2.

8.3.1 Logical Components Identification

This step of the proposed approach aims to identify high-quality components that are logical in nature, that is, it identifies the software elements that constitute a component; however, it does not deal with the physical aspects of a component that include

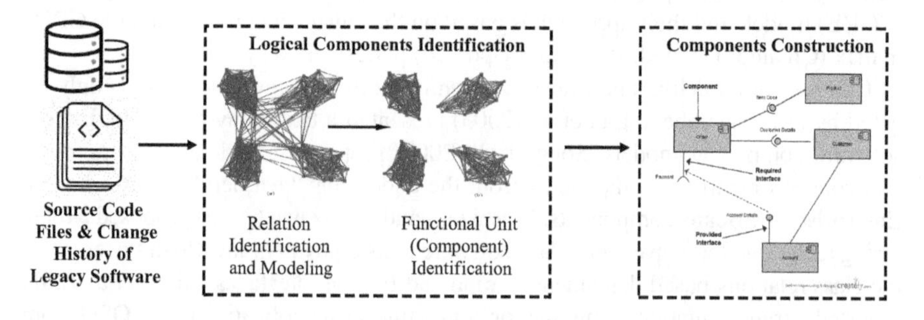

FIGURE 8.2 Proposed components identification approach from legacy system.

technological concerns, provides and requires interfaces of the component. The process of identifying logical components is a two-step process, as shown in Figure 8.2, namely, (1) relation identification and modeling, and (2) functional unit (aka component) identification. The first step aims at identifying relations (dependency-based) among different constituents of the legacy system and modeling them as a dependency graph. The second step extracts various high-quality components by performing clustering on the modeled dependency graph. These two steps are further detailed in the following subsections.

8.3.1.1 Relation Identification and Modeling

The dependency between different software system's elements (i.e., class, interface) carries out a direct relationship with the quality of components identified from it (Belguidoum and Dagnat 2007). It implies a component that possesses higher external dependencies with other components has lower quality, which restricts its reusability and vice versa. Rathee and Chhabra (2017) concludes that collective use of the following three types of dependency relation plays a key performance during the software remodularization process, namely, (1) structural relations, (2) semantic relations and (3) evolutionary relations. They are of the opinion that these three dependency relations help to measure different quality aspects of the software component that together helps in improving the overall reusability of the component. These quality aspects are identified from the different sources, such as, source code, GitHub, etc., and the extracted information is represented in the form of a dependency graph where the nodes represent different software elements and the edges between them represent the strength of different quality aspects between them.

The first quality aspect used to enhance the quality of the software component is termed as functional relatedness. The idea behind this quality aspect is that the different elements that compose a component must be functionally related, that is, they together should provide unique functionality to the outside world. In this chapter, the functional relatedness is determined using the usage pattern approach proposed by Rathee and Chhabra (2018) that helps to improve functional-based cohesion in the software system. The term usage pattern is generally used as a polysemy in software engineering (Rathee and Chhabra 2018, Saied et al. 2016, Niu et al. 2017), and it denotes the patterns available in a software system in terms of the usage for method call, member variable, library, and API, among others. In this chapter, the usage patterns are determined for different legacy software system's member variables. This is because the member variables denote the state or properties of an object in the real world. Therefore, any overlap in terms of properties represents a related entity/object in the real world. This further helps in the grouping of functionally related software elements (that together perform a more coarse-grained functionality). For a software component C that contains t different software elements defined by the set $X = \{x_1, x_2, x_3, \dots x_i, \dots, x_t\}$, the functional relatedness ($FR(C)$) is defined as shown in Equation 8.1 by averaging the individual score of functional relatedness computed at software element x_i level. Here, the function $IRD(x_i)$ represents intrafunctional relatedness degree (IRD). For a software element x_i, it is defined as the total number of usage pattern references that are made to different software elements belonging to C only. Similarly, the function $ORD(x_i)$ represents interfunctional relatedness degree

(ORD). For a software element x_i, it is defined as the total number of usage pattern references that are made to different software elements belonging to other components in the system other than *C*. For the high-quality component, the IRD of its different constituting elements should be sufficiently high compared to its ORD value.

$$FR(C) = \text{Average}\left(\sum_{i=1}^{|X|} \frac{\text{IRD}(x_i)}{\text{IRD}(x_i) + \text{ORD}(x_i)}\right) \tag{8.1}$$

The second quality aspect used for enhancing the quality of software components is termed semantic relatedness. The idea behind considering this quality aspect is to maintain semantic/conceptual coherence among different software elements of a component. Semantic relatedness helps in grouping different software elements based on the underlying concept represented by different software elements (Harispe et al. 2013). In this chapter, the semantic relatedness among different software elements is calculated based on the elementary tokens extracted from various parts of the source code of an element, namely, class/interface declaration section, member variables declaration section and function/method prototype statements. The tokens are extracted from Java source code by utilizing Java's Camel Case characteristic and breaking token to its constituent elementary token. To further enhance the quality of extracted tokens, the Porter Stemmer's algorithm is applied on extracted tokens to remove the more common morphological and inflexional endings from tokens (Dianati et al. 2014). In this chapter, the semantic relatedness of a component is computed using a well-known TF-IDF-based cosine-similarity measure (Al-Anzi and AbuZeina 2017). For a software component *C* that contains *n* different software elements defined by the set $X = \{x_1, x_2, x_3, \ldots x_i, \ldots, x_n\}$, the semantic relatedness (*SR*(*C*))is defined as shown in Equation 8.2. Here, $w_{j,i}$ is the TF-IDF weight of jth token in the ith software element.

$$SR(C) = \frac{\sum_{i=1}^{n} (w_{j,i} \; * \; w_{k,i} \; * \ldots\ldots * w_{t,i})}{\sqrt{\sum_{i=1}^{n} w_{j,i}^2} \; \sqrt{\sum_{i=1}^{n} w_{k,i}^2} \; \cdots\cdots \sqrt{\sum_{i=1}^{n} w_{t,i}^2}} \tag{8.2}$$

The third quality aspect used is the co-change relatedness and is measured using the evolutionary/change-history information associated with the software system that contains information regarding various maintenance activities performed over a period. The co-change relatedness measures the possibility of co-evolving of two or more software elements. It represents another dimension of the developer's inherent knowledge about the working of the software system. The idea behind choosing this quality dimension is that approximately 30% of dependency relations are not directly captured either by structural or semantic means, and are hidden in the evolutionary information of the software (Ajienka et al. 2018). In this chapter, the co-change relatedness is determined using **Support** and **Confidence** metrics, which are based on the association rule mining (Zimmermann et al. 2005). The support metric is defined as the proportion of transactions in which a given set of software elements are changed together and, mathematically, for a software component C that contains t different software elements as set $X = \{x_1, x_2, x_3, \ldots x_i, .., x_t\}$, as shown in Equation 8.3. The confidence metric represents a more normalized score for a given association rule and is defined in Equation 8.4. It is defined as the ratio of the proportion of transactions in which all elements of a set X are changed together with the proportion of transactions in which different software elements of X are changed independently of other elements belonging to the component C. In this chapter, the confidence metric is used to represent the co-change relatedness.

$$\text{Support}(X) = \frac{\sum_{t=1}^{|T|} \forall x_i \in T_t}{|T|}; \text{Where } x_i \in X \tag{8.3}$$

$$\text{Confidence}(C) = \frac{\text{Support}(X)}{\sum_{i=1}^{|X|} \text{Support}(x_i)}; \text{Where } x_i \in X \tag{8.4}$$

Once, the dependency relations are identified, the next step is to model these dependency relations in the form of a dependency graph. In this graph, initially, the nodes represent different software elements of the legacy software, and the edges between them represent the dependency relatedness values. Later, during the process of component identification using the proposed approach, the different nodes in this graph represent the collection of software elements.

8.3.1.2 Functional Unit (Component) Identification

The crucial and important step to identify functional unit, that is, component, is to group underlying individual software elements such that the obtained group represents a single functionality in the underlying legacy system. To perform this grouping, this chapter utilizes a nondominating-based multiobjective NSGA-II evolutionary algorithm (Deb et al. 2002). The reason behind this choice is its efficient computation complexity of $O(MN^3)$. Here, M is the number of formulated objective functions, and N is the size of the population considered for evaluation. The fitness function FF used in this chapter is based on the weighted combined use of three component quality criteria considered in Section 8.3.1 and is shown in Equation 8.5. The considered evolutionary algorithm iteratively clusters different software elements and returns logical components as different obtained clusters.

$$FF(C) = \alpha * FR(C) + \beta * SR(C) + \gamma * \text{Confidence}(C) \tag{8.5}$$

8.3.2 Components Construction

The logical components are not directly reusable and need to be converted to physical components to directly obtain reusable physical components. This step of the proposed approach aims to construct reusable physical components from the identified logical components and is achieved through transformation. This transformation needs to be implemented in two steps. First, the provides and requires interfaces of the component need to be determined. This is mandatory to make the underlying component possess two important features, namely, abstraction and encapsulation. Second, this component needs to be transformed as per the standards of the component model (e.g., JavaBeans). This transformation provides composition characteristics (i.e., a component can be connected with other components in the system) and hence reusability to the component. These two transformations are described in the following subsections.

8.3.2.1 Provides and Requires Interface Identification of a Component

The provides and requires interfaces of a component define the features/services of components that are accessible to the outside environment. Therefore, the proposed approach determines the provides and requires interfaces of a component as the name of methods belonging to a component. In the proposed approach, the provides interface is defined in terms of collection of method names that together define the services of the component, which can be accessed by other components. Similarly, the other part of the component's interface (i.e., requires) is defined in terms of collection of method names that are a part of other component's provides interface and are accessed within the component. This chapter identifies the requires interface of a component by creating a function call-based directed graph at the component level and extracting independent paths. These independent paths represent a consolidated unique functionality available in the component. In the function call-based graph, the vertex represents different method names that belong to the component, and directed edges represent the caller–callee relationship among different vertex. In our proposed approach, the first vertex becomes the part of the provides interface of the component. The reason behind considering only the first vertex in provides interface is that the rest of the vertices (methods) represent the broken (incomplete) functionality inside the component. Moreover, considering all vertices in the provides interface merely enhances understandability issues at the programmer's end rather than

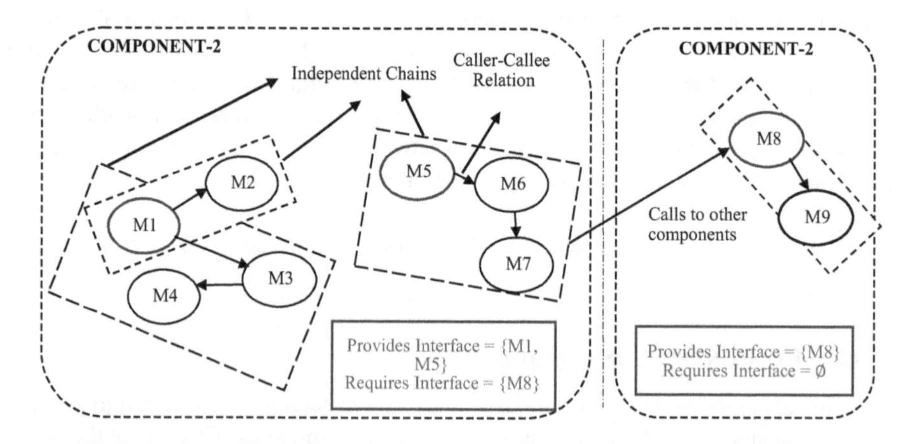

FIGURE 8.3 Function call graph and independent chains concept illustration.

doing any fruitful activity. Figure 8.3 shows a hypothetical example that illustrates the concept of provides and requires interface, function call graph and independent method chain. The considered example takes the case of two components and shows the caller–callee relationship among different methods.

8.3.2.2 JavaBeans Component Model-Based Transformation of a Component

The next step after identifying interfaces of a component is applying different transformations to the logical component as per the JavaBeans component model standard. Different transformations provide three key characteristics to the JavaBean (a component that follows JavaBeans component model standards), namely, introspection, customization, and persistence. Introspection is the ability to make the JavaBean directly understandable to third-party tools. This property helps the automatic tools in determining different services, properties and events of the JavaBean. The two types of transformations done to the logical component include (1) constructing a new software element (class) that must extend *java.beans.SimpleBeanInfo*, and (2) creating a default constructor for different constituting elements of the component (it helps in the direct instantiation of the objects of the component by third-party tools). This newly created class specifies information about different methods, properties (member variable names) and events that can be accessed to use the functionality of the underlying component. Moreover, the introspection transformation helps provide portability and reusability to the component. Further, customization provides features to the component with the help of which the component can be initialized to custom values by providing new values to its member variables. The transformation done to the logical component include creating *setter* and *getter* methods within different software elements belonging to the component. These newly created methods help in getting and initializing the values of a member variable. Their general syntax for a property named *propertyName* is shown in Equations 8.6 and 8.7.

$$\boldsymbol{SomeTypeget}PropertyName() \quad //getter\ method \tag{8.6}$$

$$\boldsymbol{SomeTypeset}PropertyName(SomeTypevalue) \quad //\ settermethod \tag{8.7}$$

Finally, the persistence transformation carried out on logical component helps in storing its current state, that is, its member variable's values and its different instance variables, to a storage device that can be retrieved later on. The necessary transformations include using the *serialization* feature of Java that provides automatic serialization of the component (JavaBean). The serialization feature in the proposed approach is provided by implementing different software elements associated with the component with the *java.io.Serializable* interface.

8.4 EXPERIMENTAL PLANNING

Experimental planning is a necessary step that includes systematic planning done for a study in order to achieve the specified objectives. Planning ensures clear and efficient availability of data required to answer different research questions of interest. Moreover, the experimental goal of the approach proposed here is to determine the feasibility of the reusable component identification process using the legacy system's source code. Therefore, this section presents the underlying technicalities of the experimental setup, such as the list of software systems studied, experimental setup considered for the NSGA-II algorithm and the list of the different research questions considered for the evaluation of the proposed approach.

8.4.1 Dataset Considered

Validation is a necessary phase to determine the feasibility of any proposed approach. This validation is performed by conducting a series of experiments on different projects. The proposed component identification technique is assessed on various open-source systems whose details are shown in Table 8.1. This table provides details such as the name of the open-source software system, its version, size in terms of the total number of classes, number of packages available in original source structure and a short description of the domain of the selected software system. The main reasons

TABLE 8.1
Summary of Considered Dataset

S.N.	Project Name	Version	No. of Classes	No. of Packages	Description
1.	JDOM[a]	2.0.6	131	15	A tool to handle XML data within Java code
2.	Apache HTTP Component Core[b]	4.4.12	468	33	An Apache project that provides low-level component support for the HTTP transport protocol
3.	Apache Commons Email[c]	1.5	20	3	API that provides simplified mailing facility
4.	ArgoUML[d]	0.35.1	835	37	A UML modeling tool
5.	JFreeChart[e]	1.5	674	33	A Java library that provides Chart draw facility
6.	JUnit[f]	4.12	183	30	A developer-side Java testing framework
7.	JavaCC[g]	7.0.5	166	6	A popular parser generator tool
8.	Logo Interpreter[h]	–	74	2	Logo language interpreter

[a] http://www.jdom.org/.
[b] http://hc.apache.org/.
[c] http://commons.apache.org/email/.
[d] http://argouml.tigris.org/.
[e] http://www.jfree.org/jfreechart/.
[f] https://github.com/junit-team/junit4.
[g] https://javacc.org/.
[h] http://naitan.free.fr/logo/.

behind considering these systems for evaluation purposes are their freely available source code, change history information and their wide applicability in the research community (Gholamshahi and Hasheminejad 2019).

8.4.2 NSGA-II Algorithm Setup

To perform multiobjective-based clustering optimization, the metaheuristic NSGA-II algorithm is used for performing the clustering of underlying software elements of a legacy system. The solution representation approach, different parameters configuration used, number of iterations of the algorithm and population size used directly affect the quality of the obtained solutions as part of the application of the NSGA-II algorithm (Li et al. 2011). The solution representation used with the NSGA-II algorithm is directly dependent on the underlying problem on which it is applied. For our current problem (identification of reusable components), it involves grouping of different independent software elements. Therefore, this chapter considers representing the feasible solution as an integer vector whose length is N, and it represents the total number of software elements belonging to a legacy system. The integer vector is randomly initialized by taking integer value in the interval $[1, N]$. This initialization helps us to randomly associate software elements to different groups that will be optimized during different iterations of the NSGA-II algorithm. The population size is considered dependent on the size of the legacy system, and is given by expression $20 * N$. Population size provides diversity to the generated population. The NSGA-II algorithm uses tournament selection and single-point crossover operators. Moreover, the probability rate for the crossover is kept as 0.8 for systems whose size $N < 100$, and 1.0 for the rest of the legacy systems. Similarly, the mutation probability is taken as $0.004 * \log_2(N)$. The NSGA-II algorithm is iterated multiple times before finally recording the result, and in each iteration, the fitness function is evaluated 25,000 times in order to sufficiently explore the search space for the algorithm. The fitness function used with the NSGA-II algorithm is already specified in Equation 8.5 and is based on the three quality characteristics of component considered in Section-8.3.1.1. Further, to minimize stochastic behavior, the NSGA-II algorithm is run 30 times on a studied software system before recording the best solution.

8.4.3 Research Questions and Evaluation Criteria

The aim of this subsection is to provide details about the evaluation criteria used for the proposed quality-focused software component identification approach. For this purpose, the following research questions were formulated:

RQ1: How efficient is the proposed component identification approach in improving the quality of a component? The aim of this formulated research question is to determine the efficiency of the proposed approach. The judgment is done according to the quality of the obtained software components. In this chapter, the overall quality of the obtained components is evaluated using the *TurboMQ* metric (Lutellier et al. 2015). This metric measures the quality in terms of inter and intradependency relations of a component.

RQ2: To what extent the proposed component construction approach preserves the semantics of the source code? The steps proposed in this chapter transform the identified logical component into its corresponding physical component that follows JavaBeans component model standards. The transformation should preserve the semantics of the underlying source code. Therefore, the formulation of this research question is carried out and is answered using the manual evaluation carried out by a team of experts.

RQ3: Does the approach stand firm in separating reusable components from a pool of legacy systems? The main aim here is to evaluate the role of different dependency relations (usage pattern based structural, semantic, and evolutionary relations) on the quality of identified components. Here, identified components are evaluated by considering different pairs of dependency relations and the evaluation is again performed using a well-known *TurboMQ* modularization quality metric.

RQ4: Do the identified interfaces follow the encapsulation principle? The interfaces of a component hide its internals and provide all access through them to the outside world. Therefore, the formulation of this research question is to determine the encapsulation potential of the identified interfaces of a logical component. This evaluation is carried out using another well-known *Abstractness* as adapted and redefined by Hamza et al. (2013).

RQ5: How does the proposed component identification approach perform against a rival approach? Such comparisons help to provide sufficient strength and validity to the proposed approach.

8.4.4 Statistical Test Performed

Statistical testing is a proven technique that helps in making a quantitative decision regarding the proposed approach. This chapter considers two statistical tests, namely, one-way ANOVA and two-tailed t-test with a confidence interval of 95% (i.e., $\propto = 5\%$). The ANOVA test is used to quantitatively study the different runs of NSGA-II algorithm and to accept/reject the null hypothesis (i.e., different obtained components in different iterations are from the distribution with the same mean and median). The two-tailed t-test is makes a quantitative decision regarding the capabilities of different dependency relations in determining the quality of the software components.

8.5 RESULT AND INTERPRETATION

The experimental results and their corresponding interpretations are provided in this section. The obtained results are presented in the form of answers to differently formulated research questions.

8.5.1 Results for RQ1 and Interpretation

RQ1 is evaluated based using the TurboMQ metric score to determine the overall quality of the studied systems. Table 8.2 shows the results for RQ1. In this table, the

TABLE 8.2
Obtained Experimental Results

				TurboMQ Metric Value		ANOVA Test
S. No.	**Project Name**	**Version**	**No. of Components**	**Before**	**After**	**p-Value**
1.	JDOM	2.0.6	21	1.45	1.95	<0.0021
2.	Apache HTTP Component Core	4.4.12	42	2.13	2.65	<0.045
3.	Apache Commons Email	1.5	5	0.55	0.69	<0.009
4.	ArgoUML	0.35.1	45	1.87	2.43	<0.0013
5.	JFreeChart	1.5	28	2.59	3.42	<0.027
6.	JUnit	4.12	35	1.23	1.57	<0.001
7.	JavaCC	7.0.5	12	2.85	3.87	<0.0035
8.	Logo Interpreter	–	6	0.86	1.17	<0.021

total number of reusable components extracted are shown, and then the quality of the components is presented followed by the ANOVA statistical test results (p-values are shown). The quality is specified in two scenarios, namely, the quality of different components/packages in the original system (here, considered as *before* case), and the quality of different components obtained as a result of applying the approach of this chapter (here, considered as *after* case).

Different logical components extracted from different legacy systems as part of experimentation are also shown in Table 8.2. Figure 8.4 plots the quality of different logical components in two scenarios, namely, *Before* and *After*. Here, it is important to note that the package structure is chosen for comparison because the proposed approach returns logical components as different clusters. From Figure 8.4, it can be directly observed that the proposed approach of this chapter results in a higher quality of identified components. The proposed approach results in significant improvement in quality that varies from a minimum of 24% to 38%. ANOVA statistical test is applied, and Table 8.2 shows the obtained p-values. The obtained p-values are significantly less than 5%, thus the null hypothesis is rejected. It means the obtained results during different iterations are statistically significant and are not merely from a distribution with a similar median. This provides a statistical strength to the proposed approach.

8.5.2 Results for RQ2 and Interpretation

This research question is evaluated based on the human knowledge of software developers. For this purpose, three independent teams of experts (each consisting of five developers of sufficient and vast experience in software development and maintenance) denoted as GP1, GP2 and GP3 were chosen by sending them personal invitations. The members of the different teams were unaware of each other to keep the obtained results unbiased. Each of these teams was presented with the original

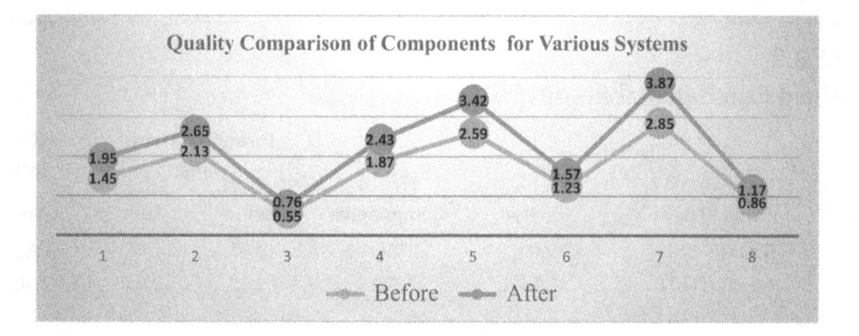

FIGURE 8.4 Quality value plot for logical components identified from legacy systems.

and transformed (after applying transformations as per the proposed approach) version of the source code of different legacy systems. Two parameters were used for the evaluation of the source code, namely, understandability (U) and functional correctness (FC). The understandability parameter measures whether the transformed code is understandable to the developers. Similarly, the FC is used to measure whether the proposed transformed code is semantically correct, that is, the meaning/function of the legacy system is the same before and after applying the transformation. As part of the evaluation, each team was asked to assign an integer value in the range of [1, 5] to each of the legacy systems with a higher number an indicator of a more understandable and functionally correct source code. Table 8.3 provides the semantic scores provided by different expert teams. The average semantic score is 4.17 and is sufficiently high to prove that based on the developer's knowledge, the proposed approach and its corresponding transformation is semantically considering the understandability and functional completeness parameters of the source code.

8.5.3 Results for RQ3 and Interpretation

This research question was designed to test the role of different dependency relations for the identification of logical components. To answer the RQ3 different dependency relations, viz. usage pattern (UP), semantic (SEM) and evolutionary (EV) were considered as independent, pair-wise, and all together (here, equal weightage is considered during different dependency relations combination). Based on this grouping, seven dependency schemes were designed, namely, UP, SEM, EV, UP+SEM, UP+EV, SEM+EV and UP+SEM+EV. The proposed approach was applied using each of the formulated schemes, and the obtained logical components were compared using TurboMQ metric score. Table 8.4 shows the obtained TurboMQ metric values against different dependency schemes. The table also shows the result of a two-tailed t-test significance test that compared the used UP+SEM+EV dependency scheme in the proposed approach against the other six formulated dependency schemes. The statistical significance test is presented using three symbols, namely, "+", "−" and "=" that represents the case where the UP+SEM+EV scheme performs better, under and equally against other schemes. From the t-test significant score mentioned in the last column of Table 8.4, it became evident that the used combined

TABLE 8.3
Results of the Manual Investigation of the Semantics by Experts

S. No.	Software Name	GP1 Group		GP2 Group		GP3 Group		Average Score
		U	FC	U	FC	U	FC	
1.	JDOM	4	4	5	5	5	5	4.67
2.	Apache HTTP Component Core	4	5	5	4	3	4	4.17
3.	Apache Commons Email	4	4	3	5	4	4	4.00
4.	ArgoUML	3	4	5	4	3	4	3.83
5.	JFreeChart	4	4	5	4	4	5	4.33
6.	JUnit	4	5	4	4	3	4	4.00
7.	JavaCC	4	4	4	4	5	5	4.33
8.	Logo Interpreter	4	4	3	5	4	4	4.00
Average Semantic Correctness Score								**4.17**

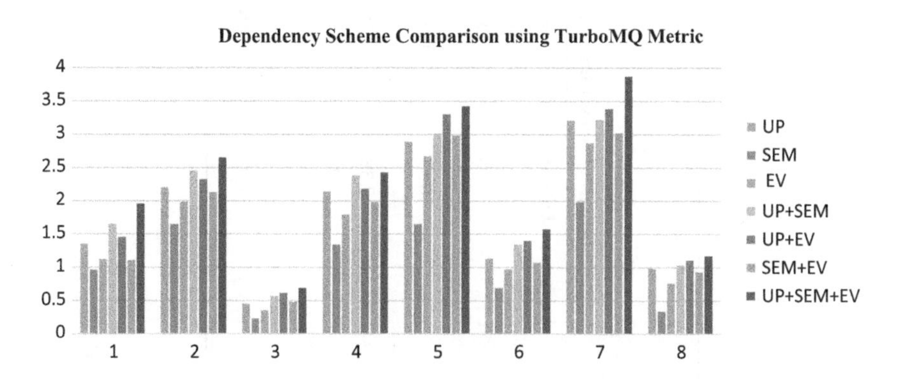

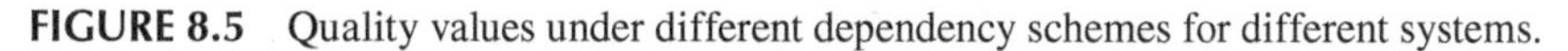

FIGURE 8.5 Quality values under different dependency schemes for different systems.

scheme (i.e., UP+SEM+EV) outperforms other dependency schemes in most cases (as specified with "+" symbol). In only a few cases, it is near equal in performance (as mentioned with "=" symbol) and in no case it underperforms.

Figure 8.5 plots different dependency schemes against various considered legacy datasets. From this figure, it is clearly visible that the proposed UP+SEM+EV scheme outperforms and gives the highest quality for the identified logical components. More noticeable is the fact that independently the semantic (SEM) relations are unable to identify quality components. Moreover, the combined dependency scheme always shows an improvement in quality.

8.5.4 Results for RQ4 and Interpretation

RQ4 evaluates the abstraction property of the component and hence determines the quality of the identified interfaces of the component. The used abstractness metric is

TABLE 8.4
Results of Different Dependency Schemes

S. No.	Software Name	TurboMQ Score Comparison							t-Test Significance Score
		UP	SEM	EV	UP+SEM	UP+EV	SEM+EV	UP+SEM +EV	
1.	JDOM	1.35	0.96	1.12	1.65	1.45	1.10	1.95	++++++
2.	Apache HTTP Component Core	2.20	1.65	1.98	2.45	2.32	2.13	2.65	++++++
3.	Apache Commons Email	0.45	0.23	0.35	0.57	0.61	0.48	0.69	++++=+
4.	ArgoUML	2.14	1.34	1.79	2.38	2.18	1.98	2.43	+++=++
5.	JFreeChart	2.89	1.65	2.67	3.01	3.30	2.98	3.42	++++=+
6.	JUnit	1.13	0.69	0.97	1.34	1.40	1.07	1.57	++++++
7.	JavaCC	3.21	1.98	2.87	3.22	3.38	3.02	3.87	++++++
8.	Logo Interpreter	0.98	0.34	0.76	1.03	1.10	0.93	1.17	++++=+

shown in Equation 8.8. This metric measures abstraction for a component by determining the total interactions with this component which are made through its provides interface (termed as $N_{\text{Interface}}$) and the number of interactions that do not involve the identified interfaces and are accessed directly and termed as N_{Direct}.

$$\text{Abstractness } (A) = \frac{N_{\text{Interface}}}{N_{\text{Interface}} + N_{\text{Direct}}} \tag{8.8}$$

Table 8.5 provides the result of the used abstractness metric values for different considered legacy systems. Figure 8.6 shows the plot of abstractness metric value in

TABLE 8.5
Component Encapsulation Verification Using Abstractness Metric

		Abstractness Metric Value		
S. No.	**Software Name**	**Before Interface Identification**	**After Interface Identification**	**Improvement Factor**
1.	JDOM	0.14	0.90	5.43
2.	Apache HTTP Component Core	0.26	0.92	2.54
3.	Apache Commons Email	0.12	0.89	6.42
4.	ArgoUML	0.15	0.87	4.80
5.	JFreeChart	0.09	0.94	9.44
6.	JUnit	0.22	0.87	2.95
7.	JavaCC	0.17	0.90	4.29
8.	Logo Interpreter	0.11	0.92	7.36

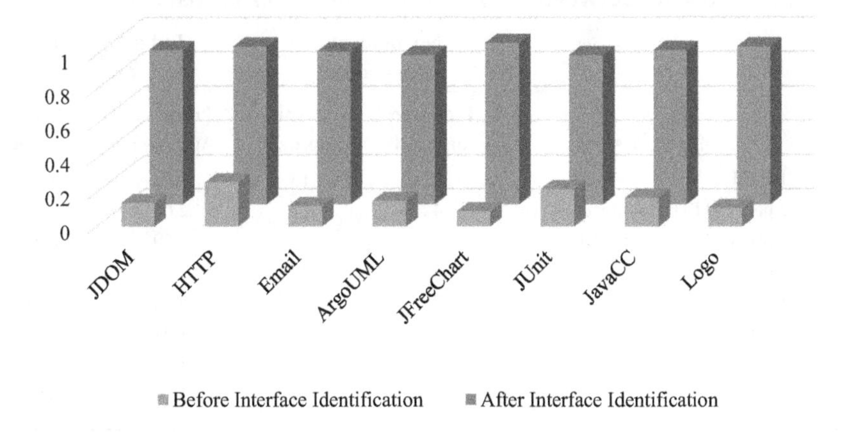

FIGURE 8.6 Abstractness metric plot for different systems.

two scenarios, namely, before and after identifying interfaces of a component. The abstractness score before identifying interfaces is sufficiently low and represents that the underlying code is directly accessed. However, after identifying interfaces, the abstractness score is quite high and represents the fact that the identified interfaces are capable of hiding underlying source code. However, it is important to note that the abstractness metric is still not 100%, denoting the fact that some accesses related to exception handling or error conditions are still not handled by the identified interfaces of the component.

8.5.5 Results for RQ5 and Interpretation

In this chapter, the considered close rival approach is the one given by Kebir et al. (2012). The considered rival approach identifies software component as a collection of classes termed as shape that consists of two main parts: the shape center (a set of classes that does not interact with classes outside the shape), and the shape interface (a set of classes that interacts with classes outside the shape). The identification is guided by the proposed fitness function designed that helps in determining the quality of a component at any time. The quality is measured in terms of specificity, autonomy and composability characteristics. Table 8.6 shows the comparison between the total number of reusable components identified using the rival and the proposed approach.

From the results shown in Table 8.6, there is no doubt that the proposed approach is capable of extracting higher quality-centric components as opposed to the rival approach. Moreover, the two approaches have the following key differences:

TABLE 8.6
Comparison Results Between the Proposed and Rival Approaches

		No. of Identified Components		TurboMQ Quality Metric	
S.N.	Software System	Proposed Approach	Rival Approach	Proposed Approach	Rival Approach
1.	JDOM	21	15	1.95	1.45
2.	Apache HTTP Component Core	42	27	2.65	2.02
3.	Apache Commons Email	5	3	0.69	0.23
4.	ArgoUML	45	36	2.43	2.23
5.	JFreeChart	28	20	3.42	3.21
6.	JUnit	35	24	1.57	1.23
7.	JavaCC	12	9	3.87	3.45
8.	Logo Interpreter	6	4	1.17	0.98

1. The rival approach does not consider the importance of semantics of a component and considers only basic cohesion and coupling metrics as a fitness function to determine the quality of different components.
2. The rival approach identifies the interfaces of a component as the group of methods belonging to only the shape interface part of the component. Moreover, the grouping is based on the heuristic correlation of method use, that is, the group which is accessed many times by other components is given priority over other groupings of methods. Here, the rival approach does not allow the participation of all the methods of a component.
3. The rival approach considered in this chapter does not carry out the transformation of the extracted components with the intention of converting logical components to corresponding physical components.

8.6 CONCLUSION AND FUTURE WORKS

Identification of reusable components is a mandatory activity for improving the success of the CBSD approach by providing components in the form of the component library. Otherwise, the required components need to be either built from scratch or purchased from third-party vendors. This chapter proposes an economical way of identifying components from the state-of-the-art legacy software systems and then constructing them as reusable components through our systematic transformation. The proposed approach consists of two phases: the first phase groups different software elements of a legacy system using the search-based NSGA-II algorithm and identifies logical components as different groups. This grouping is guided by three criteria, namely, UP-based dependency relations, token-based semantic relations and change history-based evolutionary relations, which help improve the quality of the logical component. In the second phase of the proposed approach, the logical components are transformed into physical components. This transformation is done in two steps. First, the provides and requires interfaces of the logical component are identified as the collection of different method names belonging to the component. Independent method call chains available in the method call graph of the component is used for the identification of interfaces of a component. The head vertex of such independent method chains constitutes the provides interface. The set of methods that are a part of other components and are called by the component, constitutes its requires interfaces. Second, the component is transformed according to the recommendations of JavaBeans component model standards. These transformations provide component characteristics such as introspection, customization and persistence. The proposed approach has been experimentally evaluated, and the results clearly indicate the usefulness of the approach.

In the future, testing the feasibility of multiple versions of the same software system is planned to determine their role in the quality of the identified components. The role of other kinds of dependency relations such as dynamic relations can be tested for their role in reusable component identification. An automated tool can be designed for the proposed approach that provides a visualization facility for the benefit of the user. The proposed approach can be transformed/tested to support other languages, such as C++, Ruby, among others.

REFERENCES

Abdeen, H., Sahraoui, H., Shata, O., Anquetil, N., & Ducasse, S. Towards automatically improving package structure while respecting original design decisions. In *2013 20th Working Conference on Reverse Engineering (WCRE)*. IEEE, (2013), 212–221.

Abdellatief, M., Sultan, A. B. M., Ghani, A. A. A., & Jabar, M. A. "A mapping study to investigate component-based software system metrics." *Journal of Systems and Software* 86, no. 3 (2013): 587–603.

Ajienka, N., Capiluppi, A., & Counsell, S. "An empirical study on the interplay between semantic coupling and co-change of software classes." *Empirical Software Engineering* 23, no. 3 (2018): 1791–1825.

Al-Anzi, F. S., & AbuZeina D. "Toward an enhanced Arabic text classification using cosine similarity and latent semantic indexing." *Journal of King Saud University-Computer and Information Sciences* 29, no. 2 (2017): 189–195.

Allier, S., Salah S., Sahraoui, H. & Fleurquin, R. From object-oriented applications to component-oriented applications via component-oriented architecture. In *2011 Ninth Working IEEE/IFIP Conference on Software Architecture*. IEEE, (2011), 214–223.

Alshara, Z., Seriai, A. D., Tibermacine, C., Bouziane, H. L., Dony, C., & Shatnawi, A. "Migrating large object-oriented applications into component-based ones: Instantiation and inheritance transformation." *ACM SIGPLAN Notices* 51, no. 3 (2015): 55–64.

Anthes, G. H. "Software reuse plans bring paybacks." *ComputerWorld* 27, no. 49 (1993): 73–74.

Beck, F., Melcher, J., & Weiskopf, D. Identifying modularization patterns by visual comparison of multiple hierarchies. In *2016 IEEE 24th International Conference on Program Comprehension (ICPC)*. Austin, TX, USA: IEEE, (2016), 1–10.

Belguidoum, M., & Dagnat, F. "Dependency management in software component deployment." *Electronic Notes in Theoretical Computer Science* 182 (2007): 17–32.

Bittencourt, R. A., & Guerrero, D. D. S. Comparison of graph clustering algorithms for recovering software architecture module views. In *2009 13th European Conference on Software Maintenance and Reengineering*. Austin, TX, USA: IEEE, (2009), 251–254.

Briand, L. C., Daly, J. W., & Wust, J. K. "A unified framework for coupling measurement in object-oriented systems." *IEEE Transactions on Software Engineering* 25, no. 1 (1999): 91–121.

Candela, I., Bavota, G., Russo, B., & Oliveto, R. "Using cohesion and coupling for software remodularization: Is it enough?" *ACM Transactions on Software Engineering and Methodology (TOSEM)* 25, no. 3 (2016): 24.

Chardigny, S., Seriai, A., Oussalah, M., & Tamzalit, D. Extraction of component-based architecture from object-oriented systems. In *Seventh Working IEEE/IFIP Conference on Software Architecture (WICSA 2008)*. Vancouver, BC, Canada: IEEE, (2008), 285–288.

Chhabra, J. K. "Improving modular structure of software system using structural and lexical dependency." *Information and Software Technology* 82 (2017): 96–120.

Chong, C. Y., & Lee, S. P. "Automatic clustering constraints derivation from object-oriented software using weighted complex network with graph theory analysis." *Journal of Systems and Software* 133 (2017): 28–53.

Constantinou, E., Naskos, A., Kakarontzas, G., & Stamelos, I. "Extracting reusable components: A semi-automated approach for complex structures." *Information Processing Letters* 115, no. 3 (2015): 414–417.

Corazza, A, Di Martino, S., Maggio, V., & Scanniello, G. Investigating the use of lexical information for software system clustering. In *2011 15th European Conference on Software Maintenance and Reengineering*. IEEE, (2011), 35–44.

Czibula, I. G., Czibula, G., Miholca, D. L., & Onet-Marian, Z. "An aggregated coupling measure for the analysis of object-oriented software systems." *Journal of Systems and Software* 148 (2019): 1–20.

de Oliveira Barros, M., de Almeida Farzat, F., & Travassos, G. H. "Learning from optimization: A case study with Apache Ant." *Information and Software Technology* 57 (2015): 684–704.

Deb, K., Pratap, A., Agarwal, S., & T. A. M. T. Meyarivan. "A fast and elitist multiobjective genetic algorithm: NSGA-II." *IEEE Transactions on Evolutionary Computation* 6, no. 2 (2002): 182–197.

Dianati, M. H., Sadreddini, M. H., Rasekh, A. H., Fakhrahmad, S. M., & Taghi-Zadeh, H. "Words stemming based on structural and semantic similarity." *Computer Engineering and Applications Journal* 3, no. 2 (2014): 89–99.

Ducasse, S., & Pollet, D. "Software architecture reconstruction: A process-oriented taxonomy." *IEEE Transactions on Software Engineering* 35, no. 4 (2009): 573–591.

El Hamdouni, A. E., Seriai, A. D., & Huchard, M. "Component-based architecture recovery from object oriented systems via relational concept analysis." *CLA: Concept Lattices and their Applications*, 672 (2010): 259–270.

Englander, R. *Developing JAVA Beans*. O'Reilly Media, Inc., (1997).

Erdemir, U., & Buzluca, F. "A learning-based module extraction method for object-oriented systems." *Journal of Systems and Software* 97 (2014): 156–177.

Gholamshahi, S., & Hasheminejad, S. M. H. "Software component identification and selection: A research review." *Software: Practice and Experience* 49, no. 1 (2019): 40–69.

Hamza, S., Sadou, S., & Fleurquin, R. Measuring qualities for OSGi component-based applications. In *2013 13th International Conference on Quality Software*. IEEE, (2013), 25–34.

Harispe, S., Ranwez, S. Janaqi, S., & Montmain, J. "Semantic measures for the comparison of units of language, concepts or instances from text and knowledge base analysis." *arXiv preprint arXiv:*1310.1285, (2013).

Hein, A. M. How to assess heritage systems in the early phases? Electronic format submission for AP2000. In *Proceedings of the 6th International Conference on Systems & Concurrent Engineering for Space Applications*. (2014).

Heisel, M., Santen, T., & Souquières, J. On the specification of components-the JavaBeans example. *Technical Report A02-R-025*. Nancy, France: LORIA, (2002).

Kahtan, H., Bakar, N. A., Nordin, R., & Abdulgabber, M. A. 2DCBS: A model for developing dependable component-based software. (2016).

Kebir, S., Seriai, A.-D., Chardigny, S., & Chaoui, A. Quality-centric approach for software component identification from object-oriented code. In *2012 Joint Working IEEE/IFIP Conference on Software Architecture and European Conference on Software Architecture*. IEEE, (2012), 181–190.

Lau, K. K., Ornaghi, M., & Wang, Z. A software component model and its preliminary formalisation. In *International Symposium on Formal Methods for Components and Objects*. Berlin, Heidelberg: Springer, (2005), 1–21.

Li, X., Xiao, N. Claramunt, C., & Lin, H. "Initialization strategies to enhancing the performance of genetic algorithms for the p-median problem." *Computers & Industrial Engineering* 61, no. 4 (2011): 1024–1034.

Lutellier, T., Chollak, D. Garcia, J. Tan, L. Rayside, D., Medvidovic, N., & Kroeger, R. Comparing software architecture recovery techniques using accurate dependencies. In *2015 IEEE/ACM 37th IEEE International Conference on Software Engineering*. IEEE, (2015), 69–78.

Maffort, C., Valente, M. T., Terra, R., Bigonha, M., Anquetil, N., & Hora, A. "Mining architectural violations from version history." *Empirical Software Engineering* 21, no. 3 (2016): 854–895.

Mishra, S. K., Kushwaha, D. S., & Misra, A. K. "Creating reusable software component from object-oriented legacy system through reverse engineering." *Journal of Object Technology* 8, no. 5 (2009): 133–152.

Mohammadi, S.., & Izadkhah, H. "A new algorithm for software clustering considering the knowledge of dependency between artifacts in the source code." *Information and Software Technology* 105 (2019): 252–256.

Niu, H., Keivanloo, I., & Zou, Y. "API usage pattern recommendation for software development." *Journal of Systems and Software* 129 (2017): 127–139.

Prajapati, A., & Chhabra, J. K. "TA-ABC: Two-archive artificial bee colony for multi-objective software module clustering problem." *Journal of Intelligent Systems* 27, no. 4 (2018): 619–641.

Qureshi, M. R. J., & Hussain, S. A. "A reusable software component-based development process model." *Advances in Engineering Software* 39, no. 2 (2008): 88–94.

Rathee, A.., & Chhabra, J. K. "Improving cohesion of a software system by performing usage pattern based clustering." *Procedia Computer Science* 125 (2018): 740–746.

Rathee, A., & Chhabra, J. K. "Mining reusable software components from object-oriented source code using discrete PSO and modeling them as java beans." *Information Systems Frontiers*, 78, no. 14 (2019a): 1–19.

Rathee, A., & Chhabra, J. K. "A multi-objective search based approach to identify reusable software components." *Journal of Computer Languages* 52 (2019b): 26–43.

Rathee, A., & Chhabra, J. K. "Reusability in multimedia softwares using structural and lexical dependencies." *Multimedia Tools and Applications* 78, no. 14 (2019c): 1–22.

Rathee, A., & Chhabra, J. K. "Restructuring of object-oriented software through cohesion improvement using frequent usage oatterns." *ACM SIGSOFT Software Engineering Notes* 42, no. 3 (2017): 1–8.

Rathee, A., & Chhabra, J. K. "Clustering for software remodularization by using structural, conceptual and evolutionary features." *Journal of Universal Computer Science* 24, no. 12 (2018): 1731–1757.

Saied, M. A., Ouni, A., Sahraoui, H., Kula, R. G., Inoue, K., & Lo, D. "Automated inference of software library usage patterns." *arXiv preprint arXiv* 1612.01626 (2016).

Shatnawi, A., Shatnawi, H., Saied, M. A., Shara, Z. A., Sahraoui, H., & Seriai, A. Identifying software components from object-oriented APIs based on dynamic analysis. In *Proceedings of the 26th Conference on Program Comprehension.* ACM, (2018). 189–199.

Shatnawi, A., & Seriai, A.-D. Mining reusable software components from object-oriented source code of a set of similar software. In *2013 IEEE 14th International Conference on Information Reuse \& Integration (IRI).* IEEE, (2013), 193–200.

Song, M. S., Jung, H. T., & Yang, Y. J. The analysis technique for extraction of ejb component from legacy system." In *Proceeding of 6th IASTED International Conference on Software Engineering and Applications.* Cambridge, USA: ACTA Press, (2002), 241–244.

Stafford, J. A., & Wolf, A. L. "Architecture-level dependence analysis for software systems." *International Journal of Software Engineering and Knowledge Engineering* 11, no. 04 (2001): 431–451.

Szyperski, C., Gruntz, D., & Murer, S. *Component Software: Beyond Object-Oriented Programming.* Pearson Education, 2002.

Vieira, M., Dias, M., & Richardson, D. J. Describing dependencies in component access points. In *Proceedings of the 23rd International Conference on Software Engineering, ICSE'01.* Toronto, Canada: ACM, (2001), 115–118.

Washizaki, H., & Fukazawa, Y. "A technique for automatic component extraction from object-oriented programs by refactoring." *Science of Computer Programming* 56, no. 1–2 (2005): 99–116.

Yacoub, S., Ammar, H., & Mili, A. Characterizing a software component. In *International Workshop on Component-Based Software Engineering*. (1999).

Zimmermann, T., Zeller, A., Weissgerber, P., & Diehl, S. "Mining version histories to guide software changes." *IEEE Transactions on Software Engineering* 31, no. 6 (2005): 429–445.

9 A Software Component Evaluation and Selection Approach Using Fuzzy Logic

Maushumi Lahon
Assam Engineering Institute School of Technology

Uzzal Sharma
Assam Don Bosco University

CONTENTS

9.1 INTRODUCTION

Software engineering deals with the development of software using engineering principles by a user to solve a problem. With the development of technology, challenges in the area of software development are also increasing manifold. Therefore, to cope with the challenges with respect to time, cost and technology advancement, intelligent systems are necessary to deal with the complexity associated with software development. Computational intelligence (CI) is currently used in certain areas of software development to incorporate intelligence into the decision-making process. Software architecture, software testing and requirement analysis are certain areas of software development where CI is used for better and cost-effective solutions. This chapter presents an overview of the use of CI techniques in different areas of software development with an emphasis on fuzzy logic in modeling uncertainty in software development. A component evaluation and selection approach using fuzzy logic is also proposed in this chapter.

The need for computation existed from prehistoric times as humans needed to keep an account of their physical possessions. The human need for calculating cannot be ascertained to have started from a particular date, but as human beings started to settle down and possess things, the need for keeping track of their possessions emerged. As humans evolved, analyzing and decision-making became a part of everyday life. Along with evolving humans, the development of technology gave birth to new areas of learning and research.

In the field of digital computers, different generations have evolved starting with valves, transistors, integrated circuits, microprocessors to the present generation of ultra large-scale integration. This development in technology has posed challenges to the domain of software engineering in terms of making the technology easily usable by humans. ANSI/IEEE standard 610.12-1990 defines software engineering as the application of a systematic, disciplined, quantifiable approach for development, operation and maintenance of software (Wang 2000). The domain of software engineering is evolving for more than 50 years now since the first software engineering conference held in Germany on October 7–11, 1968.

The computational power of the computer is now a tool used by humans in various domains to make intelligent decisions. Software engineering is also a domain where CI plays a role in making software development smarter. IEEE Computational Intelligence Society defines CI as neural networks, fuzzy systems and evolutionary computation, including swarm intelligence. According to literature, a clear definition of CI is yet to emerge as different methods are included and excluded under CI.

CI techniques such as neural networks, evolutionary algorithms and fuzzy logic are applied to specific software problems. Different CI paradigms are introduced in this section. All CI paradigms have their base in biological systems.

Artificial neural network (ANN) was modeled keeping the human brain as the basis. The modeling of the human neural system is termed as ANN. Learning, memorizing and performing tasks like pattern recognition and perception are a few things that form the basis of designing an artificial neural system. ANN system works basically on the same principle of gathering inputs, processing them and then generating a final output. An artificial neuron (AN) represents a biological neuron. Each AN receives a signal from the environment. An ANN is a layered network of ANs. An NN may consist of hidden layers apart from the input and output layer.

Evolutionary computation is based on the survival of the fittest principle. The natural evolution process proceeds through reproduction. In the case of an evolutionary algorithm, a population of individuals is considered and each individual is called a chromosome. Gene defines the characteristics of each chromosome, and the value of a gene is known as an allele. The objectives and constraints of a problem are represented by a fitness function, which measures the survival strength of an individual.

Some classes of evolutionary algorithms include genetic programming, evolutionary programming, evolution strategies and cultural evolution. The algorithms are used to address different types of problems.

Fuzzy logic allows an element to belong to a set with a certain degree of certainty, which is different from the traditional set theory concept of elements being a part or not a part of a set. Fuzzy logic allows approximate reasoning, as well as enables reasoning with uncertain facts to infer new facts. Fuzzy sets and fuzzy logic can be used to model common sense (Engelbrecht 2007).

9.2 LITERATURE SURVEY

The use of artificial intelligence techniques in the field of software engineering is increasing, which has given rise to conferences and publications specific to this area. In the phase of requirement engineering, fuzzy logic can be used to ascertain the requirement priorities. This, in turn, minimizes and manages the requirement conflict to some extent (Ammar et al. 2012). Bayesian networks have been used to represent the knowledge required to assess requirement specifications and predict a value which determines if the specification requires any revision (Meziane and Vadera 2010). The genetic algorithm has beenused by researchers in software architecture to find the space for hierarchical decomposition. The fitness function was based on complexity and modularity (Lutz 2001). Synthesizing architecture from requirements is also a promising area in software engineering where genetic algorithmis used (Räihä 2010).

NNs can be used to build tools for software as they have the capability to learn and adapt. Software effort estimation is an area where NN can be used for accurate estimation of effort by learning from previous estimation data. Researchers have also worked on generating software metrics using NN. In software testing, NN can be used to test software using various test criteria, test cases and metrics for coverage. Fault severity levels can also be determined using NN (Singh et al. 2009). Literature also suggests the use of Bayesian belief networks to estimate software defects by modeling the causal relationship of the software metrics in the early phases of the software development life cycle (SDLC) (Kumar and Yadav 2014).

The application of fuzzy logic in requirement analysis is by incorporating linguistic variables that are common in human communications. Requirement analysis using proper fuzzy logic can help software developers obtain correct and precise requirements. In object-oriented modeling, fuzzy logic is used to model imprecise requirements (Shinde and Sahasrabuddhe 2014). In software project management, fuzzy logic is useful because of the flexibility of inputs and outputs (Shradhanand and Jain 2007). Software reusability is another area where fuzzy logic has been used by researchers to predict the reusability from software metrics—separation of concern, coupling, cohesion and size complexity (Singh etal. 2015). Predicting software defects in each of the phases of SDLC applying fuzzy rules on the software metric data was proposed in a model. This further establishes the application of CI techniques in software engineering (Kumar and Rao 2017). Requirement prioritization is essential for developing cost-effective quality software. Fuzzy logic is applicable in prioritizing the requirements by classifying them into different levels (Mishra et al. 2016). Effort estimation is another aspect where fuzzy logic is used. In Agile software development, innovative estimation techniques are required. A review in this area shows the use of fuzzy logic for improving effort estimation by characterizing input parameters (Raslan et al. 2015). A review by Bilgaiyan et al. suggested the use of soft computing techniques to be successful in accommodating the changes introduced by users while estimating the cost of software development (Bilgaiyan et al. 2016). Fuzzy logic is also used by researchers to improve the effectiveness of software tests, thereby improving the accuracy and reliability of software development using a fuzzy weight allocation algorithm (Jiang et al. 2018).

9.3 PROBLEM IDENTIFICATION

Literature review has revealed that CI techniques have been used in the different phases of software development to cope with uncertainties in the areas of requirement analysis, requirement prioritization, fault detection, software cost and effort estimation, improving the effectiveness of testing and software project management. Hence, the relationship between software development and CI techniques are not new but there are areas of implementation which require further study and research to reap the benefits of CI techniques. In this chapter, one such area is proposed.

Different software development methodologies emphasize different strategies. Module-oriented programming, object-oriented programming, component-based software development and aspect-oriented development methodologies have their characteristics, advantages and disadvantages. Component-based development is an effective solution to minimize development costs. However, the main issue in this type of development is the selection of the appropriate component that fits into the system. The proposed life cycle models of component-based development include component selection as one of the important activities in developing the system. There are different component selection methods. In this chapter, a fuzzy logic-based component selection method is proposed to assist software developers in selecting the best component from the available components.

The technical report from the Software Engineering Institute (SEI) highlights some of the common mistakes found in the case of evaluation of COTS. These mistakes were considered to be project killers according to the report (Dorda et al. 2004). The mistakes include not performing an adequate search, no consideration to the changes in the system, no evaluation of context, no consideration of the stakeholders, that is, requirement definition does not consider the end-users and evaluation is based on marketing data without actual checking.

The report further points out the aspects that need to be considered in the process of evaluation. It consists of an understanding of the impact of COTS on the system development process, determination of the evaluation requirements, development of COTS evaluation criteria and addressing the inherent trade-offs in the evaluation process. Therefore, this report emphasizes the need and importance of the COTS evaluation process, which is discussed in this chapter.

9.4 COMPARATIVE STUDY OF AVAILABLE SOLUTION

Numerous evaluation and selection approaches have been proposed by different researchers considering different aspects and different techniques. A brief review of the techniques is given here.

OTSO (Off-the-Shelf Option): This method was proposed by Kontio et al. and is concerned only with the selection of the component and not the implementation. According to this method, COTS evaluation criteria are set and refined as the requirements change. Therefore, this process is an iterative process and the criteria for evaluation are context-dependent (Kontio 1996, Kontio et al. 1996).

PRISM (Portable Reusable and Integrated Software Module): This evaluation approach was proposed by Lichota et al. where a generic architecture was proposed, which could be used to evaluate candidate components for the system (Lichota et al. 1997). The method starts with the identification of criteria and then finding the best-fit components. Next, the components are evaluated on the basis of requirements, reusability and reliability. Finally, the integration test of the component to the system is evaluated.

STACE (Social-Technical Approach to COTS Evaluation): This approach includes socio-economic factors, that is, the nontechnical factors, such as costs, business issues, vendor performance and reliability in the evaluation and selection of COTS components (Kunda and Brooks 1999). It is an approach that covers not only the evaluation and selection aspect but also proposes techniques for the identification of the COTS components. This approach requires considerable effort for implementation.

CEP (Comparative Evaluation Process): This method was developed by the Software Productivity Consortium to evaluate and select the best-fit component among the candidate components. The method is a five-step activity, including subactivities in each step. The activities consist of scope evaluation effort (to find the expected number of COTS to be evaluated), search and screen candidate components, define evaluation criteria, evaluate component alternatives and analyze evaluation results. The decision model used to take the final decision is the simple weighted average method (Phillips and Polen 2002).

PECA (Plan, Establish, Collect, and Analyze): This approach is a cooperative effort of the Software Engineering Institute and National Research Council Canada to define a COTS software product evaluation process to plan, establish, collect, and analyze (PECA) to make sound product decisions. The activities include planning the evaluation, establishing evaluation criteria, collecting the data in regards to the candidate components and analyzing the data for new understandings that could enhance the accuracy of the evaluation process (Anderson et al. 2007).

BAREMO (Balanced Reuse Model): This model uses the concept of analytic hierarchical process (AHP) in deciding whether the component can be reused in a different software project (Tello and Gomez-Perez 2002).

CARE/SA (COTS–Aware Requirements Engineering and Software Architecting): It is an approach to match, rank and select the right component. It consists of five steps starting with defining goals, followed by matching components, ranking components, selecting components and finally negotiating the changes. Defining goals refers to specifying the requirements clearly and trying to frame a concrete requirement specification for those points the requirements defined are abstract and not specified clearly. In the next phase, a search of the existing components is conducted and the candidate components are ranked in terms of fulfilling requirement. The best matches are selected and finally a trade-off with performance-cost is made if necessary to narrow down and make the selection.

DesCOTS (Description evaluation and selection of COTS): This approach is based on the quality guidelines of the ISO/IEC9126. It is basically a software system that can be used for component evaluation and selection.

MiHOS (Mismatch handling for COTS components): This approach aims to address the mismatch aspect during and after the selection of the components. The mismatch is with respect to the requirements and the deliverables of the COTS component. The approach follows an iterative and evolutionary decision support framework and provides more than one solution for exploring and analyzing (Mohamed et al. 2007).

Table 9.1 provides an overview of the different COTS evaluation techniques stating their basic method of evaluation and the observed salient points.

The different approaches listed in the table deal with the COTS evaluation and selection process considering different variables. The objective and aim of all the approaches are to select the best match that fits the requirements of the system for which the COTS is expected to deliver.

9.5 PROPOSED SOLUTION

The approach proposed here considers the power of CI techniques to make more accurate choices in the context of COTS selection. According to this approach, the starting point is the requirement elicitation process. The requirements are functional as well as nonfunctional and need to be defined clearly before screening for the components. Once the requirements are defined, a search process for the components is initiated. Domain-specific search is important as it can considerably reduce the time required in the search process (D). The vendors in a specific area need to be identified, and the list of the components along with the services (V) is necessary for evaluation. The evaluation criteria in this approach are based on the requirements defined in the elicitation process, the functional requirements (F) and the nonfunctional requirements (NF). It is not always possible to find the components that exactly match the requirements, so the amount of effort required in resolving the mismatches (M) needs to be estimated. Hence, the variables significant for evaluating a component can be summarized as follows:

- Functional requirements (F)
- Nonfunctional requirements (NF)

TABLE 9.1
COTS Evaluation Approaches and the Identified Salient Points

Approach	Basic Method of Evaluation	Salient Points
OTSO	Searching, screening, evaluation, analysis, deployment and assessment	Detailed definition of evaluation criteria Comparison based on cost and added value
PRISM	Identification, screening, stand-alone test, integration test and field test	Has a generic component architecture Has a product evaluation process
STACE	Requirements elicitation, social-technical criteria definition, alternatives identification and evaluation	Considers both technical and nontechnical factors Use of analytic hierarchy process (AHP) for selection among alternatives
CEP	Scope evaluation effort, search and screen candidate components, define evaluation criteria, evaluate component alternatives and analyze evaluation results	Has a structured decision-making process Vendor contact and training requirements highlighted
PECA	Planning evaluation, establishing the criteria, data collection and data analysis	Creation of evaluation team Preparing a charter by defining scope, team and selection constraints
BAREMO	Based on AHP	Multicriteria decision
CARE/SA	Define goals, match components, rank components, select components and negotiate changes	COTS are represented as an aggregation of functional and nonfunctional requirements A systematic approach to match, rank and select components
DesCOTS	Based on quality guidelines ISO/IEC9126	A software system using different tools
MiHOS	Mismatch handling by finding the mismatches and their resolution, mismatch amount, technical risk, relative importance of technical goals, stakeholders and constraints	Identification of mismatches which could be zero, partial, surplus, overmatch and equivalence

- Domain-specific components (D)
- Vendors list (V)
- The effort to resolve mismatches (M)

Figure 9.1 represents the steps to be followed in the proposed selection method in the form of a flowchart.

Once the selection process is complete, there can be more than one candidate component for the system. To select the best-fit component an evaluation approach is proposed.

There could be different vendors providing the required component with various services attached to it. This may include maintenance, modification, adaptation, etc. with changing requirements. An assessment of the services provided by the vendors needs to be reviewed for a trade-off between services and estimated effort to resolve the mismatches.

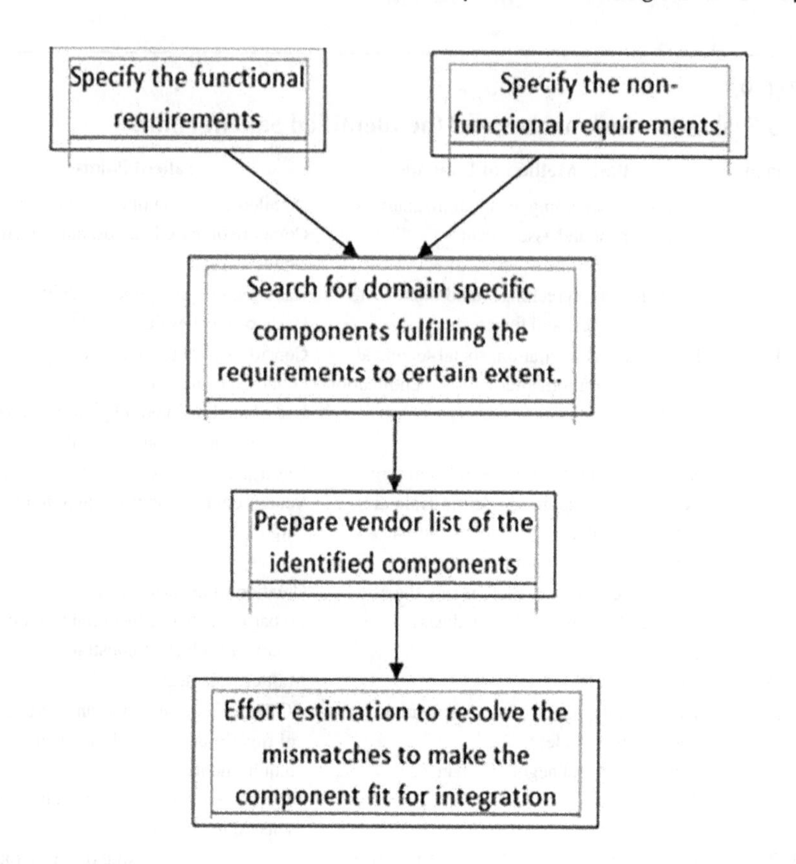

FIGURE 9.1 Steps for component selection.

Here, the use of CItechniques can be implemented to make choices with better accuracy. The proposed evaluation approach uses the fuzzy inference system (FIS) to quantify the variables determining selection.

The list of evaluation criteria may be appended depending on the type of software and its requirements. The variables are assigned a quantitative value depending on the different parameter values required to quantify the variable. There are three parameters thatare used for identifying the best-fit component in this approach:

- conformance to functional requirements (*F*)
- conformance to nonfunctional requirements (*NF*)
- effort estimation to resolve the mismatch in termsof time and cost (*E*)

To find the quantitative measure for selection, a weight is attached to each parameter. This weight can vary depending on the system requirements, for which the COTS component is to be selected. Let the selection factor be S_f which is calculated by the following equation:

$$S_f = W_1F + W_2NF + W_3E \quad (9.1)$$

Where *F* represents conformance to functional requirements, *NF* represents conformance to nonfunctional requirements and *E* represents the estimation of effort.

9.5.1 Conformance to Functional Requirements (*F*) Estimation

Consider $f_1, f_2, f_3, \ldots\ldots\ldots, f_n$ to be the different functional requirements of the system. When a component is evaluated against the functional criteria fulfillment, it is not always possible to quantify numerically. Terminologies such asslightly, moderately, considerably, etc. are used to define the level of fulfillment. Considering the different levels of fulfillment of the requirements, a quantitative value can be ascertained using a fuzzy logic-based system or FIS. An FIS is either a knowledge-based or rule-based system.The four components of FIS are fuzzifier, fuzzy rule base, fuzzy inference engine and defuzzifier. The fuzzifier uses the membership function to create the fuzzy set of the inputs. The fuzzy inference engine is the set of IF-THEN rules which is the fuzzy rule base. Defuzzifier transforms the fuzzified output to a crisp output.

The structure can be illustrated as shown in Figure 9.2.

MATLAB can be used to implement and determine the value of *F*.

In the above case of functionality determination:

The number of inputs is n

The number of output is one

The different membership functions such as triangular, trapezoidal or Gaussian can be assigned to each of the inputs and the output. The chosen membership function can then be scaled against the terminologies used in the case of the inputs and the output. For example, if we consider the terminologies defining the fulfillment of particular functional criteria be represented as poor, moderate, significant, considerable and almost, then the scaling in the range 0–1 could be 0–0.2, 0.21–0.4, 0.41–0.6, 0.61–0.8 and 0.81–1.0. Each input corresponds to one criterion (f_n) and the output is the functional requirement (*F*), which can also be scaled in similar terms or differently. Finally, a quantified value for F can be determined from the system.

9.5.2 Conformance to Non-Functional Requirements (NF)

There are many nonfunctional requirements such as reliability, interoperability, usability, etc. As defined in the case of different functional requirements, a quantitative

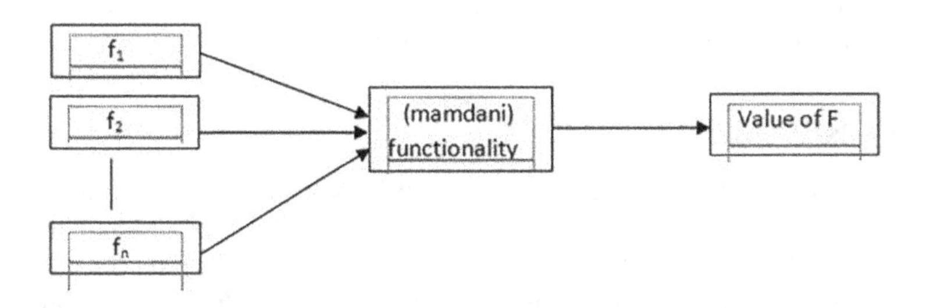

FIGURE 9.2 FIS structure for determining functionality.

measure of nonfunctional requirements can be ascertained using an FIS. As the degree of the various nonfunctional requirements cannot be quantified during evaluation and only terms like fulfilling, slightly fulfills, moderately, or considerably etc. can be used to define the degree of fulfillment of the particular nonfunctional requirement, the FIS is a very relevant system to consider the different levels of degree of conformance expressed in different words. A quantitative value for nonfunctional requirements can be ascertained following the same structure as that of functional requirements.

9.5.3 Effort Estimation (E)

The software engineering domain has different effort estimation techniques which are classified as algorithmic and nonalgorithmic. The application of CI techniques in effort estimation has been reported previously. The application of NNs for software development effort estimation, ANN in software testing effort estimation and fuzzy-based effort estimation in Agile software development have been reported in the literature.

For the purpose of the proposed approach, the effort for the adaptation of the component is to be estimated from the following three points of view:

- To fill the gap in functional requirements
- To fill the gap in nonfunctional requirements
- To fit into the system environment

This estimation can be made in terms of function points or lines of code (LOC). However, as the environment in which the component has to function are different, function point estimation is a better option than LOC.

9.6 PROS AND CONS OF SOLUTION

The proposed solution considers the responses between yes or no. As conformation to the functional or nonfunctional criteria can be expressed in terms of responses other than yes or no, better estimation of the conformation to criteria is possible. This estimation can further estimate the effort required for adaptation of the component to the environment more accurately.

The cost–benefit analysis of the solution is to be ascertained before implementation. A certified vendor list is also a necessity in this aspect to select the candidate components from the components available in the domain.

9.7 CONCLUSION

This chapter is on the use of CI technique, specifically fuzzy logic, for evaluation of a component to fit into a system developed on the component-based software development concept. There are several CI-based techniques to assist the requirement analysis process to develop better requirement specifications. The flow diagram proposed starts from requirement analysis to selection of related components, which can be done using existing approaches. The significant variables considered in the selection of components are functional requirements, nonfunctional requirements, domain-specific components, vendor list and effort to resolve mismatches. After selecting the candidate components, they are evaluated to select the best-fit component for the system. The proposed approach uses three parameters for evaluating the component: the functional specification conformance, the nonfunctional specification conformance and effort estimation. To estimate the gap between the functional and nonfunctional requirements with the selected component's capacity, the fuzzy logic approach is used to accommodate the response between yes and no. It could refer to terms such as slightly, considerably, among others. Hence, this approach is better than the existing methods. For effort estimation, existing techniques can be used, which could be by function point analysis or LOC. Function point analysis is a better option in this case as determining LOC could be difficult in this case because of the nonavailability of historical data.

9.8 FUTURE SCOPE

The proposed approach can be extended to incorporate more precise estimation by using other CI techniques to estimate the effort required for the adaptation of the component. The arbitrary weight assigned for calculating the selection factor can be further specified by considering case studies of the same domain. A quantitative estimate can be ascertained by analyzing different cases from the same domain.

REFERENCES

Ammar, H., M.S. Hamdi and W. Abdelmoez. 2012. Software Engineering Using Artificial Intelligence Techniques: Current State and Open Problems. *International Conference on Computing and Information Technology.*

Anderson, W., E. Morris, D. Smith, et al. 2007. COTS and Reusable Software Management Planning: A Template for Life-Cycle Management. *Technical Report CMU/SEI-2007-TR-011 ESC-TR-2007–011.*

Bilgaiya, S., S. Mishra and M. N. Das. 2016. A Review of Software Cost Estimation in Agile Software Development Using Soft Computing Techniques. *2nd International Conference on Computational Intelligence and Networks (CINE).* January. DOI:10.1109/CINE.2016.27.

Dorda, S.C., J.C. Dean, G. Lewis, et al. 2004. A Process for COTS Software Product Evaluation. *Technical Report CMU/SEI-2003-TR-017, ESC- TR-2003–017.* July.

Engelbrecht, A.P. 2007. *Computational Intelligence An Introduction.* John Wiley and Sons.

Jiang, D., Z. Liang and D. Huang. 2018. A Software Test Metric Method Based on Fuzzy Logic. *3rd International Conference on Information Technology and Industrial Automation (ICITIA).*

Kontio, J. 1996. A Case Study in Applying a Systematic Method for COTS Selection. *18th, International Conference on Software Engineering*, 201–209.

Kontio, J., G. Caldiera and V. R. Basili. 1996. Defining Factors, Goals and Criteria for Reusable Component Evaluation. *Proceedings of CASCON.* November, 17–28.

Kumar, C. and D. K. Yadav. 2014. Software defects estimation using metrics of early phases of software development life cycle. *International Journal of System Assurance Engineering and Management.* December. DOI: 10.1007/s13198-014-0326-2.

Kumar, T.R. and T. S. Rao. 2017. Software defects prediction based on ANN and fuzzy logic using software metrics. *International Journal of Applied Engineering Research*, 12(19): 8509–8517. March, ISSN 0973-4562.

Kunda, D. and L. Brooks. 1999. Applying Social-Technical Approach for Cots Selection. *Proceedings of 4th UKAIS Conference*, University of York, McGraw Hill.

Lichota, R.W., R.L. Vesprini and B. Swanson. 1997. PRISM Product Examination Process for Component Based Development. *Proceedings Fifth International Symposium on Assessment of Software Tools and Technologies, IEEE.*

Lutz, R. 2001. Evolving good hierarchical decompositions of complex systems. *Journal of Systems Architecture*, 47: 613–634. DOI: 10.1016/S1383-7621(01)00019-4.

Meziane, F. and S. Vadera. 2010. Artificial Intelligence in Software Engineering Current Developments and Future Prospects. In *Artificial Intelligence Applications for Improved Software Engineering Development: New Prospects*. IGI Global. Chapter 14, 278–299, DOI: 10.4018/978-1-60566-758-4.ch014.

Mishra, N., M.A. Khanum and K. Agrawal. 2016. Approach to Prioritize the Requirements Using Fuzzy Logic. *ACEIT Conference Proceedings.*

Mohamed, A., G. Ruhe and A. Eberlein. 2007. *MiHOS: An Approach to Support Handling the Mismatches Between System Requirements and COTS Products*. Springer-Verlag London Limited. DOI: 10.1007/s00766-007-0041-5.

Phillips, B.C. and S. M. Polen. 2002. Add decision analysis to your COTS selection process. *CROSSTALK The Journal of Defense Software Engineering*, 21–25, January.

Räihä, O. 2010. A survey on search-based software design. *Computer Science Review*, 4: 203–249.

Raslan, A.T., N.R. Darwish and H. A. Hefny. 2015. Towards a fuzzy based framework for effort estimation in Agile software development. *International Journal of Computer Science and Information Security*, 13(1): 37–45, January.

Shinde, M. and D. Sahasrabuddhe. 2014. Software requirement scaling using fuzzy logic. *International Journal of Application or Innovation in Engineering & Management*, 3 (3): 564–568, March.

Shradhanand, K. A. and S. Jain. 2007. Use of fuzzy logic in software development. *Issues in Information Systems*, VIII(2): 238–244.

Singh, P. K., O.P. Sangwan, A. P. Singh, et al. 2015. A framework for assessing the software reusability using fuzzy logic approach for aspect oriented software. *International Journal of Information Technology and Computer Science*. January, DOI: 10.5815/ijitcs.2015.02.02.

Singh, Y., P.K. Bhatia, A. Kaur et al. 2009. Application of Neural Networks in Software Engineering: A Review. *Conference Paper in Communications in Computer and Information Science,* March. DOI: 10.1007/978-3-642-00405-6_17.

Tello, A.L. and A. Gomez-Pérez. 2002. BAREMO: How to choose the appropriate software component using the analytic hierarchy process. Proceedings of the 14th international conference on *Software Engineering and Knowledge Engineering*, (SEKE'02) July, 781–788.

Wang, J.U. 2000. Towards Component-Based Software Engineering *Proceedings of the Eighth Annual Consortium on Computing in Small Colleges Rocky Mountain Conference*, 177–189.

BIBLIOGRAPHY

1. Rich, E., K. Knight and S.B. Nair. 2009. *Artificial Intelligence.* Tata McGraw Hill Education Private Limited.

2. Giarratano, J.C. and G.D. Riley. 2008. *Expert Systems: Principles and Programming.* Cengage Learning India Private Limited.
3. Alpaydin, E. 2010. *Introduction to Machine Learning.* PHI Learning Private Limited.
4. Sommerville, I. 2004. *Software Engineering.* Pearson Education India.
5. Pressman, R.S. 2017. *Software Engineering: A Practitioner's Approach.* McGraw Hill Education.

10 Smart Predictive Analysis for Testing Message-Passing Applications

Mohamed Elwakil
University of Cincinnati Blue Ash College

CONTENTS

10.1 INTRODUCTION

Given a run-time model of a program and a set of safety properties, predictive analysis is used to determine whether there exists a feasible interleaving of the events in the run-time model that violates any of the functional correctness properties of the program. Predictive analysis combines features from testing and model-checking.

Similar to testing, the input program is run and its behavior is observed. We can view an application execution trace as an abstract model for the input program that is analyzed exhaustively to detect potential errors; akin to the model-checking technique (Clarke, Grumberg and Peled 1999), predictive analysis is not as comprehensive as model-checking, but it is efficient and scalable (Savage, et al. 1997). Predictive analysis has been used successfully to detect concurrency errors in multithreaded programs (Wang, et al. 2009; Sen and Agha 2005).

Applying smart predictive analysis involves four steps: (1) obtain a run-time model of an input program, (2) enumerate all feasible execution interleavings that can be derived from the run-time model, (3) check every interleaving to determine if it violates any functional correctness property and (4) inspect violating interleavings to verify that they can occur in the input program. Explicitly enumerating all feasible execution scenarios is a bottleneck that hinders the performance of smart predictive analysis. We have developed a trace-based SMT-driven predictive analysis technique that reduces the process of explicitly enumerating and checking all interleavings to constraint solving that employs off-the-shelf SMT (Bouton, et al. 2009) (Satisfiability Modulo Theories) solvers.

The set of rules that govern mapping the constructs of a trace to the quantifier-free first-order logic (QF-FOL) formula are called an *encoding*. We have devised two different encodings for MCAPI (Brehmer 2013) program traces: state-based encoding and events-based encoding. Devising the rules that map messaging and other programming constructs (i.e., the trace constructs) to SMT constructs is a daunting research venture because of two reasons: (1) first, there is a huge semantic gap between the trace constructs and the SMT constructs. Even with such a gap, the logic formula should accurately mimic the trace constructs. Otherwise, the formula will be incorrect as it will allow execution scenarios that can never happen. (2) Second, at the same time, replicating the tiny details of the trace constructs will produce an inefficient formula that does not scale well with the size of the input trace. Thus, the mapping rules must strike a balance between accuracy and efficiency.

Also, we have developed the mzPredictor tool that implements our trace-based SMT-driven predictive analysis technique. mzPredictor is a push-button solution that takes as input an MCAPI program source code and produces as output a report that describes a specific execution scenario that violates the functional correctness of the input program. In the next section, we describe the workflow of the mzPredictor and its components.

10.2 THE MZPREDICTOR WORKFLOW

The mzPredictor is a push-button solution that takes as input an MCAPI program source code and produces as output a report that describes a specific execution scenario that violates the functional correctness of the input program. Figure 10.1 shows the mzPredictor workflow and highlights its three components: mzInstrumenter, mzEncoder and mzReporter.

The mzInstrumenter instruments a program source code by adding extra code that monitors the program execution and emits *events* during run-time. An event indicates the execution of a particular program statement that is being monitored.

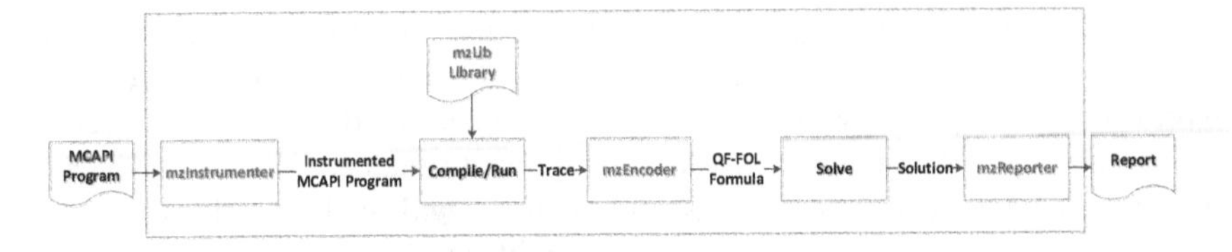

FIGURE 10.1 The mzPredictor workflow.

The amalgamation of captured events constitutes a trace of the input program. The resulting trace is accumulated in memory and dumped to a file before the termination of the instrumented program. The monitoring is lightweight to minimize probe-effect (Gait 1986) but broad enough to allow generating an accurate symbolic representation of the trace. Additionally, the trace maintains the path condition of each event. An event path condition is the logical conjunction of all conditions necessary for executing the corresponding statement. Section 10.3 details the instrumentation process and describes the trace grammar.

The mzEncoder translates the captured trace to a QF-FOL formula that consists of abstract variables and constraints over the values of the abstract variables. The translation is performed according to a set of rules (i.e., an encoding) that map the trace constructs to SMT constructs that are restricted to the background theories supported by the target SMT solver. We have devised two encodings. We call the first one the *state-based* encoding. Preliminary experiments with the state-based encoding were encouraging. However, more thorough experiments revealed that it does not scale well when the number of potential races increases. The state-based encoding uses arrays to mimic MCAPI communication channels. Solving formulas with arrays is very time consuming using current SMT solvers (de Moura and Bjøner 2008). Hence, we developed another encoding that does not use arrays, and we call it the *order-based* encoding. The state-based encoding is described in Section 10.4 and the order-based encoding is described in Section 10.5.

We use Yices (Dutertre and Moura 2006) to solve the formula generated by the mzEncoder. If the formula is solvable (i.e., satisfiable), then Yices outputs a solution that assigns values to the abstract variables.

The solution produced by Yices embodies an execution scenario that violates the program's functional correctness. Finally, the mzReporter (described in Section 10.6) transforms this solution into a report that describes a concrete execution scenario showing how functional correctness of the program is violated.

10.3 THE mzINSTRUMENTER

The mzInstrumenter uses Eclipse CDT (Foundation 2013) to instrument an input program by inserting *logging* functions calls at specific locations in the code. These functions communicate with a monitoring engine (the mzLib library in Figure 10.1). The mzLib library maintains the in-memory trace and saves it to the disk when the program execution ends.

The mzInstrumenter performs the following tasks: (1) add auxiliary code before node starting/exiting points to initialize/tear down the monitoring engine; (2) add function calls that register program variables, including *for* loops iteration variables; (3) since SMT solvers cannot handle prefix/postfix operators, prefix and postfix expressions are replaced with equivalent expressions that do not contain prefix or postfix operators; (4) add code that logs assignment statements including *for* loops update statements; (5) replace an assert statement with code that logs the assert invariant; (6) replace MCAPI routines calls with other calls that generate corresponding events and invoke the original MCAPI calls; (7) extend the parameters list of each added logging function to include the path condition and the line number of the corresponding program statement.

The path condition (Elwakil and Yang 2010) of a program statement is the logical conjunction of all conditions in the program path leading to the statement. For instance, if the statement is in the else-part of an *if* statement, which, in turn, is inside a *while* loop, then the path condition will be the logical conjunction of the negation of the *if* statement condition and the *while* loop condition.

After the instrumentation is complete, the instrumented program is compiled and run. When the instrumented program terminates, it dumps a file that contains an execution trace (i.e., the logged events).

10.3.1 The Trace Grammar

The execution trace of an MCAPI program running on N nodes will have N subtraces, a subtrace for each node. Let $\mathcal{R} = \{\mathcal{T}_1,\ldots,\mathcal{T}_n\}$ be the trace of an MCAPI program with N nodes. $\mathcal{T}_n$ is the subtrace produced by node N_n and it consists of a sequence of events $\mathcal{T}_n = T_{n,1}\ldots T_{n,|\mathcal{T}_n|}$. Each subtrace $\mathcal{T}_n$ is associated with a set of instance variables: $\mathcal{L}_n = \{L_{n,1},\ldots, L_{n,|\mathcal{L}_n|}\}$, and a set of end-points employed in this node: $\varepsilon\mathcal{P}_n = \{EP_{n,1},\ldots,EP_{n,|\varepsilon\mathcal{P}_n|}\}$.

An event $T_{n,x} \in \mathcal{T}_n$ is a tuple $\langle y, Guard, Action\rangle$, such that n is a node identifier, x is the order of $T_{n,x}$ appearance in $\mathcal{T}_n$, y is the order of proceeding event in $\mathcal{T}_n$, *Guard* is a condition that must be true for this event to take place and *Action* is an atomic computation that corresponds to an executed statement in the program. If $T_{n,x}$ is the last event in $\mathcal{T}_n$, then y is set to the special marker value $\perp$. *Guard* is the logical conjunction of all conditions in the program path leading to the statement that produced the event. For instance, if the statement that produces $T_{n,x}$ is in the then-part of an *if* statement, which, in turn, is inside a *while* loop, then *Guard* will be the logical conjunction of the *if* statement condition and the *while* loop condition. *Action* can be any of the following types:

- *Assign* (v, exp) corresponds to an assignment statement that assigns *exp* to v. $v \in \mathcal{L}_n$ is a variable and *exp* is an expression over $\mathcal{L}_n$.
- *Send*$(src, dest, exp)$ corresponds to an `mcapi_msg_send` statement that sends a message from *src* to *dest*, which contains *exp*. $src \in \varepsilon\mathcal{P}_n$ and $dest \in \varepsilon\mathcal{P}_m$ are the source and destination end-points, respectively. *exp* is an expression over $\mathcal{L}_n$.
- Similarly, *Send_i* $(src, dest, exp, req)$ corresponds to an `mcapi_msg_send_i` statement where $req \in \mathcal{L}_n$ is a request variable.

TABLE 10.1
Trace Example

$\mathcal{T}_1$		$\mathcal{T}_3$	
$T_{1,1}$	Assign(Msg,1)	$T_{3,1}$	Assign(Msg,10)
$T_{1,2}$	Send_i(EP1,EP2,Msg,r0)	$T_{3,2}$	Send(EP3,EP2,Msg)
$T_{1,3}$	Send_i(EP1,EP4,Msg,r1)	$T_{3,3}$	Send(EP3,EP4,Msg)
$T_{1,4}$	Wait(r0)	$\mathcal{T}_4$	
$T_{1,5}$	Wait(r1)	$T_{4,1}$	Assign(U,0)
$\mathcal{T}_2$		$T_{4,2}$	Assign(W,0)
$T_{2,1}$	Assign(X,0)	$T_{4,3}$	Assign(O,0)
$T_{2,2}$	Assign(Y,0)	$T_{4,4}$	Recv(EP4, U)
$T_{2,3}$	Assign(Z,0)	$T_{4,5}$	Assert(U>0)
$T_{2,4}$	Recv_i(EP2,X,r2)	$T_{4,6}$	Recv(EP4, W)
$T_{2,5}$	Recv_i(EP2,Y,r3)	$T_{4,7}$	Recv(EP4, O)
$T_{2,6}$	Wait(r2)		
$T_{2,7}$	Wait(r3)		
$T_{2,8}$	Assign(Z, X-Y)		
$T_{2,9}$	Send(EP2, EP4, Z)		

- *Recv* (*recv*,*v*) corresponds to an `mcapi _ msg _ recv` statement that receives a message at the receiving end-point $recv \in \mathcal{EP}_n$. The message contents are assigned to variable $v \in \mathcal{L}_n$.
- Similarly, *Recv_i* (*recv*,*v*,*req*) corresponds to an `mcapi _ msg _ recv _ i` statement where $req \in \mathcal{L}_n$ is a request variable.
- *Wait* (*req*) corresponds to an `mcapi _ wait` statement that waits for the completion of an asynchronous action whose status is tracked with request variable $req \in \mathcal{L}_n$.
- *Assert* (*exp*) corresponds to an `assert` statement with the expression *exp*. *exp* is a Boolean expression over $\mathcal{L}_n$.

Table 10.1 shows the events of a trace of a program with four nodes. The first and third nodes send two numbers to the second and fourth nodes, respectively. Node 2 computes the difference between the gathered numbers and transmits that difference to node 4. Node 4 expects to receive three numbers, with the first number being more significant than zero. We will be using this trace as an ongoing example.

10.4 STATE-BASED ENCODING

A symbolic program execution is a limited sequence of program states (Ben-Ari 1990), with each state a projection from abstract variables to concrete values. The first state of an execution maps all variables to initial values. Each consecutive state is derived by carrying out a single instruction. In the case of concurrent programs consisting of simultaneously running components, instructions are chosen from each component instructions sequence, one at a time and are interleaved in some total order. The last state of a symbolic execution (Lee, et al. 2009) is that immediately following the execution of the last available instruction from the sequence of any component.

The state-based encoding produces a symbolic replica of the events in the trace, and solving the formula produced by this encoding is equivalent to symbolically executing the trace with different orderings of the trace events with the goal of finding an ordering that violates any of the functional correctness properties of the program. In other words, when attempting to solve the formula produced by the state-based encoding, the SMT solver *generates* and *simulates* the execution of all possible valid interleavings of a symbolic model of the original program. The formula consists of two parts: abstract variables and constraints.

10.4.1 The Abstract Variables

The abstract variables correspond to the variables that appear in the program in addition to auxiliary variables. In a trace that has B events, we will generate $B + 1$ symbolic states ($\mathcal{S}^0, \mathcal{S}^1, \ldots, \mathcal{S}^i, \ldots, \mathcal{S}^B$), such that $\mathcal{S}^0$ is the initial state, and $\mathcal{S}^i$ is the state on the ith point of time; which occurs after executing the ith event's action. A state $\mathcal{S}^i$ is a snapshot of every abstract variable accessible at point of time i. To model the $B + 1$ state, we introduce $B + 1$ instances of the variables appearing in the trace. For instance, $L^i_{n,x}$ represents the version of the variable $L_{n,x}$ at the ith time step.

In any moment of time, one of the actions, labeled the *pending action*, at a specific node, labeled as the *active node*, will be executed in a symbolic fashion. The variable NS^i (node's selector) denotes the node that will be activated at moment i. The value of NS^i is randomly chosen by the SMT solver. The determination of the value of NS^i is not completely arbitrary but is affected by *scheduling constraints.* The impending action of a node N_n is defined by the counter variable of the node NC_n. The range of NC_n is $\{1 \ldots |T_n|, \perp\}$. $NC_n = x$ signifies that the impending action in the node with identifier N_n will be $T_{n,x}$. $NC_n = \perp$ implies that all actions in node N_n have been executed in a symbolic manner.

The storage areas acting as network buffers for end-points are abstracted as queues. For a destination end-point $EP_{n,x}$, that receives one or more messages, we create a corresponding queue $Q_{n,x}$. Q_n is all abstract queues needed to encode the target end-points in node N_n. A queue $Q_{n,x}$ is abstracted as a list with two variables $head_{n,x}$ and $tail_{n,x}$ that encapsulate the start and end positions of the array. The MCAPI specification employs asynchronous *send* and *receive* calls: *mcapi_msg_send_i*, and *mcapi_msg_recv_i*. The run-time utilizes signal objects to follow the condition of an asynchronous call. Such calls instigate an operation (i.e., a *send* or a *receive*), initializes a signal object to pending, and exits instantly. The conclusion of an asynchronous invocation may be checked by calling the synchronous `wait` call and providing it with the signal object created by the asynchronous call. The `wait` invocation exits after the asynchronous call has finished. An asynchronous *send* is considered concluded when the message has arrived to the MCAPI buffer. An asynchronous *receive* is concluded when the run-time completes loading the relevant message from the buffer. A signal object will be represented as an abstract variable that can assume one of these values: PND, CMP and NLL.

10.4.2 The Constraints

The state-based encoding is composed of four constraints: the initialization constraint ($\mathcal{F}_{init}$), the actions constraint ($\mathcal{F}_{acts}$), the scheduling constraint ($\mathcal{F}_{sched}$) and the property constraint ($\mathcal{F}_{prp}$).

10.4.2.1 The Initial Constraint ($\mathcal{F}_{init}$)

The initialization constraint ($\mathcal{F}_{init}$) sets the variables values at point 0. All the counters of the node are set to the value 1. $\mathcal{F}_{init}$ is expressed as:

$$\wedge_{n=1}^{|\mathcal{R}|}\left(\left(NC_n^0 = 1\right) \wedge \left(\wedge_{v=1}^{|\mathcal{L}_n|} L_{n,v}^0 = iv_{n,v}\right) \wedge \left(\wedge_{q=1}^{|\mathcal{Q}_n|} head_{n,q}^0 = tail_{n,q}^0 = 0\right)\right) \quad (10.1)$$

where $iv_{n,v}$ is the first value for the variable $L_{n,v}$. We consider that the signal objects used in a node N_n are in the instance variables ($\mathcal{L}_n$) of a node, and we initialize them to the special value NLL.

10.4.2.2 The Action Constraint ($\mathcal{F}_{acts}$)

The actions constraint ($\mathcal{F}_{acts}$) represents the outcome of executing a pending action. It is a logical conjunction of B formulas ($\mathcal{F}_{acts} = \wedge_{i=1}^{B} \mathcal{F}_{act^i}$), such that $\mathcal{F}_{act^i}$ encodes the action selected to be executed at the ith point of time. The $\mathcal{F}_{act^i}$ constraint relies on the action category as described below:

For an event $T_{n,x} = \langle y, Guard, Action\rangle$ whose $Action = Assign(v,exp)$, the corresponding constraint formula is:

$$\left(NS^i = n \wedge NC_n^i = x \wedge Guard\right) \rightarrow \left(NC_n^{i+1} = y \wedge v^{i+1} = exp^i \wedge \delta\left(\{v^i\}\right)\right) \quad (10.2)$$

Formula 10.2 expresses the fact that at point of time i, if node N_n is designated as active ($NS^i = n$), node N_n's counter will be equivalent to x ($NC_n^i = x$) and the *Guard* holds true, then the counter in the point $i+1$ is set to the order of the next action ($NC_n^{i+1} = y$), the value of variable v in the point of time $i+1$ is set to the expression *Expr* at point of time i ($v^{i+1} = exp^i$) and all instance variables except v and all heads and tails of the buffer queues will be set in point $i+1$ to identical values to what they had in point of time i ($\delta(\{v^i\})$) .

For an event $T_{n,x} = \langle y, Guard, Action\rangle$ whose $Action = Send(src,dest,exp)$, the corresponding constraint formula is:

$$\left(NS^i = n \wedge NC_n^i = x \wedge Guard\right) \rightarrow$$

$$\left(NC_{nid}^{i+1} = y \wedge Q_{dest}\left[tail_{dest}^i\right] = exp^i \wedge tail_{dest}^{i+1} = tail_{dest}^i + 1 \wedge \delta\left(\{tail_{dest}^i\}\right)\right) \quad (10.3)$$

Formula 10.3 encodes the fact that at moment i, if node N_n is designated as active ($NS^i = n$), node N_n's instruction pointer is equivalent to x ($NC_n^i = x$), and *Guard* holds true, then the counter in the moment $i+1$ is set to y ($NC_n^{i+1} = y$), the dispatched expression is added to the queue at the target end-point queue ($Q_{dest}[tail^i_{dest}] = exp^i \wedge tail^{i+1}_{dest} = tail^i_{dest} + 1$) and all instance variables and every head and tail value of a queue except $tail^i_{dest}$ should have at instant $i+1$ identical values to what they had in the preceding point of time $i(\delta(\{tail^i_{dest}\}))$.

For an action $T_{n,x} = \langle y, Guard, Action \rangle$ whose $Action = Recv(recv,v)$, the corresponding constraint formula is:

$$\left(NS^i = n \wedge NC^i_n = x \wedge Guard\right) \rightarrow \begin{pmatrix} NC^{i+1}_n = y \wedge v^{i+1} = Q_{recv}\left[\, head^i_{recv} \,\right] \wedge \\ head^{i+1}_{recv} = head^i_{recv} + 1 \wedge \delta\left(\left\{v^i, head^i_{recv}\right\}\right) \end{pmatrix} \quad (10.4)$$

Formula 10.4 abstracts the fact that, at point of time i, if node N_n is designated as active ($NS^i = n$), node N_n's instruction pointer is equal to x ($NC^i_n = x$), and the *Guard* holds true, then the counter variable in the point of time $i+1$ is set to y ($NC^{i+1}_n = y$), the received value is removed from the queue at the destination end-point queue ($v^{i+1} = Q_{recv}[head^i_{recv}] \wedge head^{i+1}_{recv} = head^i_{recv} + 1$) and all instance variables except v^i and every head and tail value of a queue but $= head^i_{recv}$ will be set in $i+1$ to identical values to what they had at the earlier point of time i ($\delta(\{v^i, head^i_{recv}\})$).

For an event $T_{n,x} = \langle y, Guard, Action \rangle$ whose $Action = Send\_i(src,dest,exp,req)$, the corresponding constraint formula is:

$$\left(NS^i = n \wedge NC^i_n = x \wedge Guard\right) \rightarrow$$

$$\begin{pmatrix} NC^{i+1}_n = y \wedge Q_{dest}\left[\, tail^i_{dest} \,\right] = exp^i \wedge tail^{i+1}_{dest} = tail^i_{dest} + 1 \wedge req^{i+1} \\ = PND \wedge \delta\left(\left\{tail^i_{dest}, req^i\right\}\right) \end{pmatrix} \quad (10.5)$$

Formula 10.5 represents the fact that, at point of time i, if node N_n is designated as active ($NS^i = n$), node N_n's instruction pointer is equivalent to x ($NC^i_n = x$), and *Guard* holds true, then the counter variable in the point of time $i+1$ is set to y ($NC^{i+1}_n = y$), the sent expression is added to the queue of the destination end-point ($Q_{dest}[tail^i_{dest}] = exp^i \wedge tail^{i+1}_{dest} = tail^i_{dest} + 1$), the value of the signal object variable is set to pending ($req^{i+1} = PND$), and all instance variables but req^i and every head and tail value of a queue but $tail^i_{dest}$ will be set in point $i+1$ to identical values to what they had at the previous point of time i ($\delta(\{tail^i_{dest}, req^i\})$).

For an event $T_{n,x} = \langle y, Guard, Action \rangle$ whose $Action = Recv\_i(recv,v,req)$, the corresponding constraint formula is the logical conjunction of F6 and F7:

$$\left(NS^i = n \wedge NC^i_n = x \wedge Guard \wedge head^i_{recv} = tail^i_{recv}\right) \rightarrow$$
$$\left(NC^{i+1}_n = y \wedge req^{i+1} = PND \wedge \delta\left(\left\{req^i\right\}\right)\right) \quad (10.6)$$

$$\left(NS^i = n \wedge NC_n^i = x \wedge Guard \wedge head_{recv}^i \neq tail_{recv}^i\right) \rightarrow$$
$$\left(\begin{array}{l} NC_n^{i+1} = y \wedge v^{i+1} = Q_{recv}\left[\; head_{recv}^i \right] \wedge head_{recv}^{i+1} = head_{recv}^i + 1 \wedge req^{i+1} \\ \quad = CMP \wedge \delta\left(\left\{req^i, head_{recv}^i, v^i\right\}\right) \end{array}\right) \tag{10.7}$$

Formula 10.6 indicates that, at point of time i, if node N_n is designated as active ($NS^i = n$), node N_n's instruction pointer is equal to x ($NC_n^i = x$), and the *Guard* holds true, and the receiving end-point queue is empty ($head_{recv}^i = tail_{recv}^i$), then the counter variable in the point of time $i+1$ is set to y. ($NC_n^{i+1} = y$), the request variable is set to pending ($req^{i+1} = PND$), and all instance variables but req^i and every head and tail value of a queue will be set in point $i+1$ to identical values to what they had at point of time i ($\delta(\{req^i\})$).

Formula 10.7 expresses the fact that, at point of time i, if node N_n is designated as active ($NS^i = n$), node N_n's instruction pointer is equal to x ($NC_n^i = x$), the *Guard* holds true, and the receiving end-point queue is not empty ($head_{recv}^i \neq tail_{recv}^i$), then the counter variable in the point of time $i+1$ is set to y ($NC_n^{i+1} = y$), the request variable is set to complete ($req^{i+1} = CMP$), the received value is dequeued from the receiving end-point queue ($v^{i+1} = Q_{recv}[head_{recv}^i] \wedge head_{recv}^{i+1} = head_{recv}^i + 1$) and all instance variables but v^i and req^i and every head and tail value of a queue but $head_{recv}^i$ will be set in point $i+1$ to identical values to what they had at point of time i ($\delta(\{v^i, head_{recv}^i, req^i\})$).

For an event $T_{n,x} = \langle y, Guard, Action \rangle$ whose $Action = Wait(req)$, such that req is associated with an asynchronous send, the corresponding constraint formula is:

$$\left(NS^i = n \wedge NC_n^i = x \wedge Guard\right) \rightarrow \left(NC_n^{i+1} = y \wedge req^{i+1} = CMP \wedge \delta\left(\left\{req^i\right\}\right)\right) \tag{10.8}$$

Formula 10.8 asserts that, at point of time i, if node N_n is designated as active ($NS^i = n$), node N_n's instruction pointer is equal to x ($NC_n^i = x$), and the *Guard* holds true, then the counter variable in the point of time $i+1$ is set to y ($NC_n^{i+1} = y$), the request variable is set to complete ($req^{i+1} = CMP$), and all instance variables but req^i and every head and tail value of a queue will be set in point $i+1$ to identical values to what they had at point of time i ($\delta(\{req^i\})$).

For an event $T_{n,x} = \langle y, Guard, Action \rangle$ whose $Action = Wait(req)$, such that req is associated with an asynchronous receive action with $Action = Recv\_i(recv, v, req)$, the corresponding constraint formula is the logical conjunction of F9 and F10:

$$\left(NS^i = n \wedge NC_n^i = x \wedge Guard \wedge req^i = PND\right) \rightarrow$$
$$\left(\begin{array}{l} NC_n^{i+1} = y \wedge v^{i+1} = Q_{recv}\left[\; head_{recv}^i \right] \wedge head_{recv}^{i+1} \\ \quad = head_{recv}^i + 1 \wedge req^{i+1} = CMP \wedge \delta\left(\left\{req^i, head_{recv}^i, v^i\right\}\right) \end{array}\right) \tag{10.9}$$

Formula 10.9 encodes the fact that, at point of time i, if node N_n is designated as active ($NS^i = n$), node N_n's instruction pointer is equal to x ($NC_n^i = x$), the *Guard* holds true, and the request is pending ($req^i = PND$), then he instruction pointer in the point of time $i+1$ is set to ($NC_n^{i+1} = y$), the request variable is set to complete ($req^{i+1} = CMP$), the received value is dequeued from the receiving end-point queue ($v^{i+1} = Q_{cv}[head_{recv}^i] \wedge head_{recv}^{i+1} 1$) and all instance variables but v^i and req^i and every head and tail value of a queue but $head_{recv}^i$ will be set in point $i+1$ to identical values to what they had at point of time i $(\delta(\{v^i, head_{recv}^i, req^i\}))$.

Formula 10.10 abstracts that at point of time i, if node N_n is designated as active ($NS^i = n$), node N_n's instruction pointer is equal to x ($NC_n^i = x$), the *Guard* holds true, and the request is complete ($req^i = CMP$), then the counter variable in the point of time $i+1$ is set to the order of the next action ($NC_n^{i+1} = y$), and all instance variables and every head and tail value of a queue will be set in point $i+1$ to identical values to what they had at point of time i ($\delta(\phi)$).

10.4.2.3 The Scheduling Constraint ($\mathcal{F}_{sched}$)

The scheduling constraint ($\mathcal{F}_{sched}$) is the logical conjunction of B constraints ($\mathcal{F}_{sched} = \wedge_{i=1}^{B} \mathcal{F}_{sched^i}$). Each $\mathcal{F}_{sched^i}$ constraint is composed of four elements to guarantee that:

1. A node that has executed all its actions can't become an active node:

$$\bigwedge_{n=1}^{|\mathcal{R}|}\left[\left(NC_n^i = \perp\right) \rightarrow \left(NS^i \neq n\right)\right] \tag{10.11}$$

2. If a node becomes inactive, its variable values will remain unchanged:

$$\bigwedge_{n=1}^{|\mathcal{R}|}\left[\left(NS^i \neq n\right) \rightarrow \left(\bigwedge_{v=1}^{|\mathcal{L}_n|} L_{n,v}^{i+1} = L_{n,v}^i\right) \wedge \left(NC_n^{i+1} = NC_n^i\right)\right] \tag{10.12}$$

3. A node with a pending synchronous receive action whose destination queue is empty, will not be an active node:

$$\bigwedge_{n=1}^{|\mathcal{R}|}\left[\left(NC_n^i = x \wedge Action_{n,x} = \mathrm{Recv}(recv, v) \wedge head_{recv}^i = tail_{recv}^i\right) \rightarrow \left(NS^i \neq n\right)\right] \tag{10.13}$$

4. A node with a pending wait action of an asynchronous receive whose destination queue is empty, will not be an active node:

$$\bigwedge_{n=1}^{|\mathcal{R}|}\left[\left(\begin{array}{l} NC_n^i = x \wedge Action_{n,x} = Wait(req) \wedge \exists\, \mathrm{Recv\_i}(recv, v, req) \wedge head_{recv}^i \\ = tail_{recv}^i \rightarrow \left(NS^i \neq n\right) \end{array}\right)\right] \tag{10.14}$$

10.4.2.4 The Property Constraint ($\mathcal{F}_{prp}$)

The property constraint ($\mathcal{F}_{prp}$) captures the functional correctness properties of the program and is derived from events whose *Action* = *Assert*(*exp*) as follows:

$$\bigwedge_{n=1}^{|\mathcal{R}|}\left[NC_n^i = x \wedge Action_{n,x} = Assert(\exp) \wedge Guard \rightarrow Expr^i\right] \quad (10.15)$$

The complete formula sent to the SMT solver is constructed as follows:

$$\mathcal{F}_{\mathcal{R}} = \mathcal{F}_{init} \wedge \mathcal{F}_{acts} \wedge \mathcal{F}_{sched} \wedge \neg\mathcal{F}_{prp} \quad (10.16)$$

If Formula 10.16 is satisfiable, then its solution represents an execution scenario that violates one or more of the functional correctness constraints encoded by $\mathcal{F}_{prp}$.

10.4.3 Experimental Results

Because no public MCAPI benchmarks are available, we conducted our experiments using applications and benchmarks that we developed. We used a machine equipped with a Core 2 Duo running at 1.4 GHz and 4 GB of RAM.

In the first iteration of experiments, we applied the state-based encoding on the traces of nine MCAPI applications with varying features. The generated formulas were then input to Yices. We assessed the applicability of the state-based encoding by measuring the time and memory needed by Yices to solve the formula generated by the encoding. The run-time and memory consumed by Yices to solve the formulas are depicted in Figure 10.2. The *x*-axis reports the number of events in the input trace. These results looked promising as all the formulas in the experiment could be solved in a fraction of a second and required a small amount of memory.

However, applying the state-based encoding on the traces of more complicated programs revealed that it doesn't scale well. In the second iteration of experiments, we used the following benchmarks:

1. **Binary Tree (BT)**: it is a group of ten applications that generate networks of connected nodes which range in size between 3 and 21 nodes. The children of a node send to their parent node messages to establish a binary tree

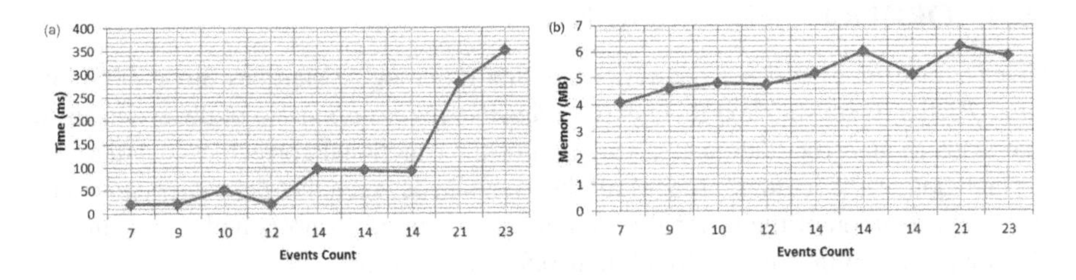

FIGURE 10.2 Results of the first iteration of experiments on state-based encoding.

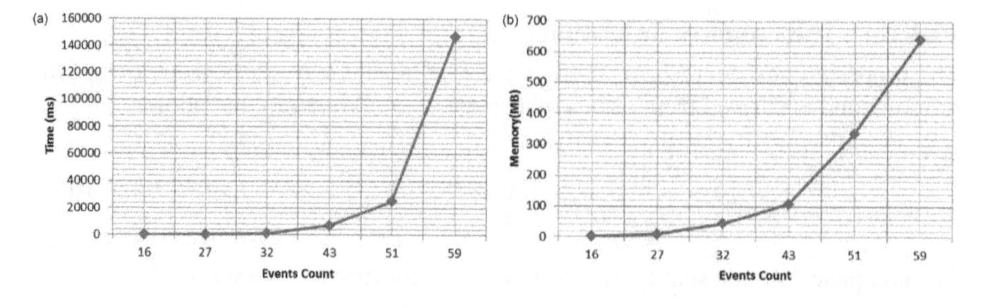

FIGURE 10.3 Results of applying state-based encoding on the BT benchmark.

pattern in which messages move from the leaves toward the root node. The smallest network has three nodes and delivers four messages. The largest network has 21 nodes and delivers 31 messages. This benchmark communication pattern is known as the master/slave pattern.

2. **Complete Graph (CG)**: This benchmark consists of ten applications that produce networks of connected nodes with sizes starting at two nodes up to 11 nodes. All nodes exchange messages with all other node.. The amount of the dispatched messages ranges from two (for a two-node graph) to 11 (for an 11-node structure). This benchmark communication pattern is called the all-to-all pattern.
3. **Ten-Nodes (TN)**: This benchmark consists of ten applications that create networks of nodes with a static size of ten nodes. However, the number of messages communicated between the nodes rises monotonically. The number of messages exchanged is between ten and 100.

Figure 10.3 depicts the results of applying the state-based encoding on the traces of the BT benchmark programs. Only the results of the first six programs are available. Starting from the seventh program, Yices exceeded a time-out of 5 minutes. The programs of the other benchmarks started timing-out at the third program for the CG benchmark and the second program for the TN benchmark.

The reason behind the poor performance of the state-based encoding is rooted in using arrays to encode the MCAPI run-time buffers. Available SMT solvers cannot handle formulas with arrays efficiently. In the next section, we present the order-based encoding, which is array-free.

10.5 ORDER-BASED ENCODING

Given a trace $\mathcal{R}$, we create a quantifier-free first-order logic (Marques-Silva 2008) formula $\mathcal{F}_{\mathcal{R}}$ that is satisfiable if there exists a *feasible permutation* $\mathcal{P}_{\mathcal{R}}$ of the events in $\mathcal{R}$ that leads to an error state (e.g., a violation of the functional correctness). A feasible permutation is a strict total order of all the events in $\mathcal{R}$, such that this order can occur in a real execution of the original program. The abstract variables and constraints of $\mathcal{F}_{\mathcal{R}}$ are presented in Sections 10.5.1 and 10.5.2, respectively.

10.5.1 The Abstract Variables

In this encoding, there are two types of abstract variables: (1) for every event $T \in \mathcal{R}$, there is an abstract variable $\mathcal{O}_T$ that reflects the *order* of carrying out T in $\mathcal{P}_R$. (2) For every action that assigns a new value to an instance variable L, we create a new abstract variable for L (e.g., L_i corresponds to the value of the variable L after the ith assignment). The values of these abstract variables record the history of the values of L. This is like the SSA form (Lee, Padua and Midkiff 1999). While the SSA form requires φ functions to handle the effect of branches, we need not have φ functions because, in a trace, all branching decisions have already been made. We add two dummy variables $\mathcal{O}_{First}$ and $\mathcal{O}_{Last}$, such that $\mathcal{O}_{First}$ is the first event in $\mathcal{P}_R$ and $\mathcal{O}_{Last}$ is the last event in $\mathcal{P}_R$. The values assigned to these abstract variables are governed by constraints that are crafted to ensure that $\mathcal{P}_R$ is a feasible permutation. An abstract variable $\mathcal{O}_{i,j}$ represents the order of the event $T_{i,j}$.

Table 10.2 shows the abstract variables that are needed for encoding the trace in Table 10.1. An abstract variable $\mathcal{O}_{i,j}$ represents the order of the event $T_{i,j}$. An abstract variable T_iV_j corresponds to the value of the variable V at subtrace T_i after being assigned a new value for the jth time.

TABLE 10.2
Abstract Variables of the Trace in Table 10.1

$\mathcal{T}_1$	$\mathcal{T}_2$	$\mathcal{T}_3$	$\mathcal{T}_4$
$\mathcal{O}_{1,1}$	$\mathcal{O}_{2,1}$	$\mathcal{O}_{3,1}$	$\mathcal{O}_{4,1}$
$\mathcal{O}_{1,2}$	$\mathcal{O}_{2,2}$	$\mathcal{O}_{3,2}$	$\mathcal{O}_{4,2}$
$\mathcal{O}_{1,3}$	$\mathcal{O}_{2,3}$	$\mathcal{O}_{3,3}$	$\mathcal{O}_{4,3}$
$\mathcal{O}_{1,4}$	$\mathcal{O}_{2,4}$	T_3Msg_1	$\mathcal{O}_{4,4}$
$\mathcal{O}_{1,5}$	$\mathcal{O}_{2,5}$		$\mathcal{O}_{4,5}$
T_1Msg_1	$\mathcal{O}_{2,6}$		$\mathcal{O}_{4,6}$
	$\mathcal{O}_{2,7}$		$\mathcal{O}_{4,7}$
	$\mathcal{O}_{2,8}$		T_4U_1
	$\mathcal{O}_{2,9}$		T_4W_1
	T_2X_1		T_4O_1
	T_2Y_1		T_4U_2
	T_2Z_1		T_4W_2
	T_2X_2		T_4O_2
	T_2Y_2		
	T_2Z_2		

10.5.2 The Constraints

The order-based formula has four constraints: the order constraint ($\mathcal{F}_{order}$), the assignment constraint ($\mathcal{F}_{asgn}$), the receive constraint ($\mathcal{F}_{recv}$) and the property constraint ($\mathcal{F}_{prp}$). The $\mathcal{F}_{\mathcal{R}}$ formula is the logical conjunction of these four constraints:

$$\mathcal{F}_{\mathcal{R}} := \mathcal{F}_{order} \wedge \mathcal{F}_{asgn} \wedge \mathcal{F}_{recv} \wedge \neg \mathcal{F}_{prp} \quad (10.17)$$

10.5.2.1 The Order Constraint ($\mathcal{F}_{order}$)

$\mathcal{F}_{order}$ ensures that in $\mathcal{P}_{R}$, no two events are assigned the same ordering and that every two events $T_{i,x}$ and $T_{i,y}$, such that $x < y$ (i.e., event $T_{i,x}$ appears in the trace before event $T_{i,y}$) will be assigned orderings $\mathcal{O}_{i,x}$ and $\mathcal{O}_{i,y}$, such that $\mathcal{O}_{i,x} < \mathcal{O}_{i,y}$. $\mathcal{F}_{order}$ is constructed using the algorithm Construct_FOrder in Figure 10.4.

10.5.2.2 The Assignment Constraint ($\mathcal{F}_{asgn}$)

$\mathcal{F}_{asgn}$ encodes events with assignment actions. $\mathcal{F}_{asgn}$ is initially set to true. For every event $T_{i,x}$ whose action is *Assign(v,exp)*:

$$\mathcal{F}_{asgn} := \mathcal{F}_{asgn} \wedge \left(S(v) = S(exp) \wedge S(Guard)\right) \quad (10.18)$$

```
1   F_order := true
2   for i=1 to n
3       F_order := F_order ∧ (O_First < O_T{i,1})
4       for j=1 to |T_i|
5           if (j<|T_i|) then  F_order := F_order ∧ (O_T{i,j} < O_T{i,j+1})
6           for k=i+1 to n
7               for l=1 to |T_k|
8               F_order := F_order ∧ (O_T{i,j} ≠ O_T{k,l})
9               end-for
10          end-for
11      end-for
12      F_order := F_order ∧ (O_Last > O_T{i,j})
13  end-for
```

FIGURE 10.4 The Construct_FOrder() algorithm.

where *S(v)*, *S(exp)*, and *S(Guard)* replace the program variables with the corresponding symbolic ones.

10.5.2.3 The Receive Constraint ($\mathcal{F}_{recv}$)

$\mathcal{F}_{recv}$ encodes the events with an action that is either a synchronous receive or a wait of an asynchronous receive. To facilitate describing the $\mathcal{F}_{recv}$ constraint, we use the following notations:

For every event $T_{i,x}$ whose *Action* is either *Send(src,dest,exp)* or *Send_i (src,dest,exp,req)*:

- $DestEP(T_{i,x}) = dest$
- $Exp(T_{i,x}) = exp$
- $SOrder(T_{i,x})$ is the order of $T_{i,x}$ with respect to other events in $\mathcal{T}_i$ whose actions are either *Send(src,dest,exp)* or *Send_i(src,dest,exp,req)* and have the same destination end-point as $T_{i,x}$.

For every event $T_{i,x}$ whose *Action* is either *Recv(recv,v)* or *Wait(req)* such that *Wait(req)* is associated with an asynchronous receive action *Recv_i(recv,v,req)*:

- $RecvEP(T_{i,x}) = recv$
- $Var(T_{i,x}) = v$
- $ROrder(T_{i,x})$ is the order of $T_{i,x}$ with respect to other events in $\mathcal{T}_i$ whose actions are either *Recv(recv,v)* or *Wait(req)* such that *Wait(req)* is associated with an asynchronous receive action *Recv_i(recv,v,req)* and have the same receiving end-point as $T_{i,x}$
- $\mathcal{S}_{i,x}$ is the set of events whose actions are either *Send(src,dest,exp)* or *Send_i(src,dest,exp,req)* and can potentially match with the receive action of $T_{i,x}$. $\mathcal{S}_{i,x}$ is defined as: $\mathcal{S}_{i,x} = \{T_{j,y} \mid DestEP(T_{j,y}) = RecvEP(T_{i,x}) \land ROrder(T_{i,x}) \geq SOrder(T_{j,y})\}$. We call $\mathcal{S}_{i,x}$, the *set of potential sender events* of $T_{i,x}$.
- $\mathcal{P}_{i,x}$ is the set of events whose actions are (1) either *Recv(recv,v)* or *Wait(req)* such that *Wait(req)* is associated with an asynchronous receive *Recv_i(recv,v,req)* (2) precede $T_{i,x}$ in $\mathcal{T}_i$, and (3) have the same receiving end-point as $T_{i,x}$. $\mathcal{P}_{i,x}$ is defined as: $\mathcal{P}_{i,x} = \{T_{i,y} \mid ROrder(T_{i,y}) < ROrder(T_{i,x})\}$. We call $\mathcal{P}_{i,x}$, the *set of related preceding receiving events* of $T_{i,x}$.

$\mathcal{F}_{recv}$ is initially set to true. For an event $T_{i,x}$ whose action is either *Recv(recv,v)* or *Wait(req)* such that *Wait(req)* is associated with an asynchronous receive *Recv_i(recv,v,req)*:

$$\begin{aligned}\mathcal{F}_{recv} := \mathcal{F}_{recv} \land \lor_{s \in \mathcal{S}_{i,x}} (S(Var(T_{i,x})) \\ = S(Exp(s)) \land S(Guard) \land CON^{s}_{T_{i,x}} \land \land_{p \in \mathcal{P}_{i,x}} \neg CON^{s}_{p})\end{aligned} \quad (10.19)$$

$$CON^{s}_{r} = (O_s < O_r) \land \land_{n \in \mathcal{S}_r \land - n \neq s} ((O_n < O_s) \lor (O_r < O_n)) \quad (10.20)$$

CON_{r}^{s} encodes the *conditions* needed for matching an event s with a send action to an event r with a receive action. These conditions are (1) s must precede r ($O_s < O_r$), and (2) for every event n, such that $n \in S_r \land n \neq s$, then either n is before s, or r is before ($\bigwedge_{n \in S_r \land n \neq s} ((O_n < O_s) \lor (O_r < O_n))$).

According to Formula 10.19 the receive action of $T_{i,x}$ will be matched with the event s when the conditions for this matching are satisfied ($CON_{T_{i,x}}^{s}$), and when all the conditions needed for matching s with any event in $P_{i,x}$ are not satisfiable ($(\bigwedge_{p \in P_{i,x}} \neg CON_{p}^{s})$).

For instance, the part of $\mathcal{F}_{recv}$ that corresponds to the event $T_{4,4}$ is the disjunction of Formulas 10.21–10.23. Formulas 10.21–10.23 match the receive action at event $T_{4,4}$ with the send action at events $T_{1,3}$, $T_{2,9}$ and $T_{3,3}$, respectively, and encodes the necessary conditions. Only one formula of these three formulas will be satisfied.

$$(T_4U_2 = T_1Msg_1 \land (O_{1,3} < O_{4,4} \land (((O_{3,3} < O_{1,3}) \lor (O_{4,4} < O_{3,3})) \land (O_{2,9} < O_{1,3}) \lor (O_{4,4} < O_{2,9})))) \quad (10.21)$$

$$(T_4U_2 = T_2Z_1 \land (O_{2,9} < O_{4,4} \land (((O_{3,3} < O_{2,9}) \lor (O_{4,4} < O_{3,3})) \land (O_{1,3} < O_{2,9}) \lor (O_{4,4} < O_{2,9})))) \quad (10.22)$$

$$(T_4U_2 = T_3Msg_1 \land (O_{3,3} < O_{4,4} \land (((O_{2,9} < O_{3,3}) \lor (O_{4,4} < O_{2,9})) \land (O_{1,3} < O_{3,3}) \lor (O_{4,4} < O_{3,3})))) \quad (10.23)$$

Intuitively, $\mathcal{F}_{recv}$ matches an event $s \in S_r$ with *one* event r provided that s has not been matched with any event $p \in P_r$, and s can occur before r. The effect of a matching is assigning the valuation of the expression sent by s to the variable of r.

10.5.2.4 The Property Constraint ($\mathcal{F}_{prp}$)

The property constraint ($\mathcal{F}_{prp}$) captures the functional correctness properties of the program and is derived from events with *Assert(exp)* actions as follows:

$\mathcal{F}_{prp}$ is initially set to true. For every event $T_{i,x}$ whose action as *Assert(exp)*:

$$\mathcal{F}_{prp} := \mathcal{F}_{prp} \land (S(exp) \land S(Guard)) \quad (10.24)$$

After the formula $\mathcal{F}_{\mathcal{R}}$ has been constructed, it is passed to an SMT solver. If $\mathcal{F}_{\mathcal{R}}$ is satisfiable, then the SMT solver will produce a solution that assigns a value for every O_T variable that indicates the order of carrying out the event T in the permutation $P_{\mathcal{R}}$.

10.5.3 Experimental Results

Figure 10.5 depicts the results of applying the order-based encoding on the traces of the BT benchmark programs.

Comparing Figures 10.3 and 10.5 shows clearly that the order-based encoding scales better than the state-based encoding.

Figures 10.6 and 10.7 depict the results of applying the order-based encoding on the traces of the CG and TN benchmarks, respectively.

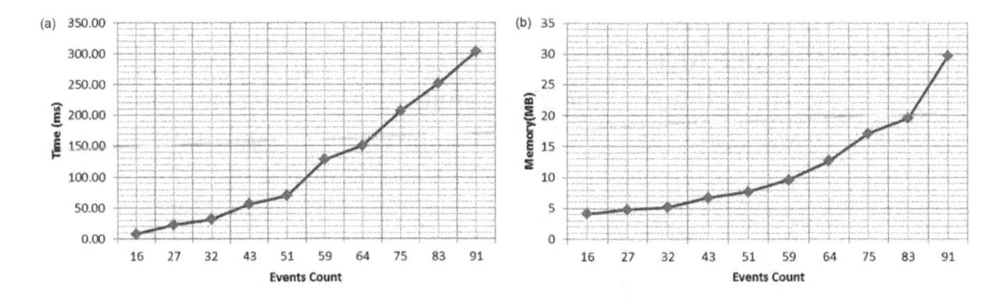

FIGURE 10.5 Results of applying order-based encoding on the BT benchmark.

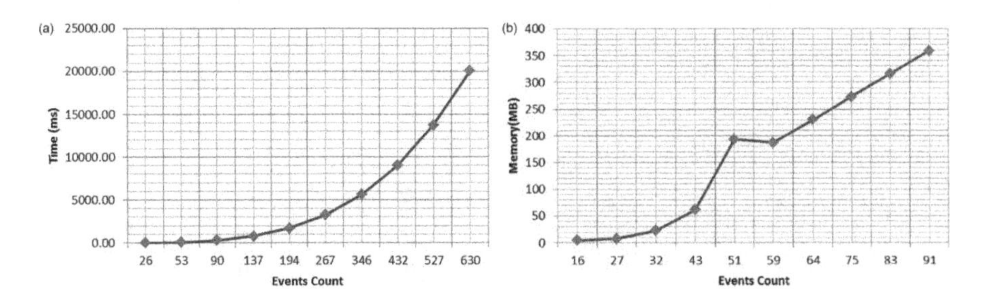

FIGURE 10.6 Results of applying order-based encoding on the CG benchmark.

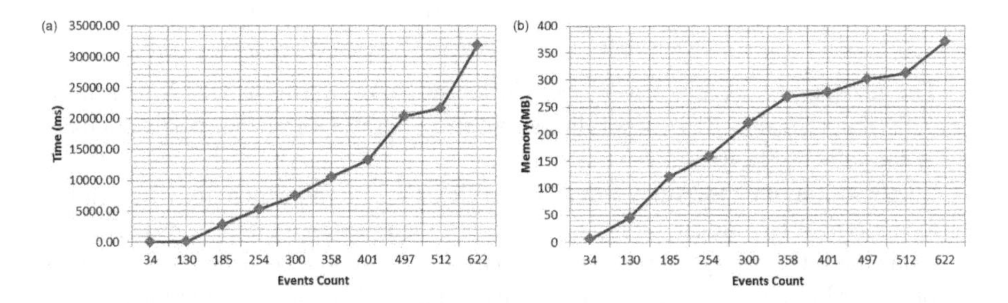

FIGURE 10.7 Results of applying order-based encoding on the TN benchmark.

The graphs in Figures 10.5–10.7 show that the order-based encoding exhibits better scalability than the step-based encoding in terms of time and memory usage.

10.6 REPORTING A VIOLATING SCENARIO

Table 10.3 shows the solution produced by Yices for the formula that corresponds to the trace in Table 10.1 for the order-based encoding.

According to the trace in Table 10.1, any of the events ($T_{1,3}$, $T_{3,3}$, and $T_{2,9}$) whose actions are send actions can match with the event $T_{4,4}$ whose action is a receive action. In Table 10.3 $O_{2,9} < O_{4,4}$, $O_{1,3} > O_{4,4}$ and $O_{3,3} > O_{4,4}$ indicating that $T_{2,9}$ is the event that will be matched with $T_{4,4}$.

TABLE 10.3
The Solution of the $\mathcal{F}_{\mathcal{R}}$ Formula

Variable	Value	Variable	Value
$\mathcal{O}_{1,1}$	1	T_2Y_2	10
$\mathcal{O}_{1,2}$	7	T_2Z_2	–9
$\mathcal{O}_{1,3}$	19	$\mathcal{O}_{3,1}$	11
$\mathcal{O}_{1,4}$	21	$\mathcal{O}_{3,2}$	13
$\mathcal{O}_{1,5}$	22	$\mathcal{O}_{3,3}$	23
T_1Msg_1	1	T_3Msg_1	10
$\mathcal{O}_{2,1}$	2	$\mathcal{O}_{4,1}$	9
$\mathcal{O}_{2,2}$	3	$\mathcal{O}_{4,2}$	10
$\mathcal{O}_{2,3}$	4	$\mathcal{O}_{4,3}$	12
$\mathcal{O}_{2,4}$	5	$\mathcal{O}_{4,4}$	17
$\mathcal{O}_{2,5}$	6	$\mathcal{O}_{4,5}$	18
$\mathcal{O}_{2,6}$	8	$\mathcal{O}_{4,6}$	20
$\mathcal{O}_{2,7}$	14	$\mathcal{O}_{4,7}$	24
$\mathcal{O}_{2,8}$	15	T_4U_1	0
$\mathcal{O}_{2,9}$	16	T_4W_1	0
T_2X_1	0	T_4O_1	0
T_2Y_1	0	T_4U_2	–9
T_2Z_1	0	T_4W_2	1
T_2X_2	1	T_4O_2	10

The mzReporter receives a Yices solution as input and translates it to a user-friendly report. Table 10.4 shows the output of the mzReporter that corresponds to the solution in Table 10.3.

The first column shows the order of carrying the actions which are listed in the third column. When the action is a `msg _ recv` or a wait of a `recv _ i`, the fourth column shows the change in the variable that is receiving the arriving message. This report describes a concrete execution scenario that leads to an assertion failure. The actual report includes a fifth column showing the line number of each action in the input source code.

10.7 CONCLUSION

We have presented a methodology and a proof-of-concept implementation for predicting run-time failures in MCAPI programs by symbolically enumerating and examining all permutations of the events in an execution trace. We have developed two different encodings that are used to translate an execution trace to a quantifier-free first-order logic formula.

Our approach is practical, scalable and sound. It is practical because it is fully automated and does not require manual annotating of the source code. Our experimental results show that our approach is scalable, particularly when using the second encoding. It is sound as no spurious execution scenarios are reported.

TABLE 10.4
mzReporter Report

Order	Node	Action	Side Effect
1	1	`Msg=1`	
2	2	`X=0`	
3	2	`Y=0`	
4	2	`Z=0`	
5	2	`mcapi_msg_recv_i(EP2,X,r2)`	
6	2	`mcapi_msg_recv_i(EP2,Y,r3)`	
7	1	`mcapi_ msg_i(EP1,EP2,Msg,r0)`	
8	2	`mcapi_wait(r2)`	X=1
9	4	`U=0`	
10	4	`W=0`	
11	3	`Msg=10`	
12	4	`O=0`	
13	3	`mcapi_msg_ send(EP3,EP2,Msg)`	
14	2	`Wait(r3)`	Y=10
15	2	`Z=X-Y`	Z=–9
16	2	`mcapi_msg_send(EP2, EP4,Z)`	
17	4	`mcapi_msg_recv(EP4,U)`	U=–9
18	4	`Assert (U>0)`	Failure!

Constructing the symbolic formula from the trace instead of the program leads to two limitations: (1) our approach lacks completeness as some bugs in the input program may escape detection because we are considering a trace of the program, rather than the program itself, (2) also, our approach cannot handle nondeterminism arising from routines such as `mcapi _ wait _ any`, `mcapi _ test` and random numbers generation.

These limitations could be overcome by combining elements from the input program with the trace. Another solution is to repeatedly execute the program to obtain a different trace, and hence, increasing the analysis coverage; however, there are no guarantees that different executions will produce different traces. Moreover, MCAPI programs are vulnerable to deadlocks (Sharma, et al. 2009). We plan to explore how to encode an MCAPI program trace such that it is possible to predict deadlocks.

REFERENCES

Ben-Ari, M. 1990. *Principles of Concurrent and Distributed Programming*. Upper Saddle River, NJ: Prentice Hall.

Bouton, T., D. de Oliveira, D. Déharbe, and P. Fontaine. 2009. \sf veriT: an open, trustable and efficient SMT-solver. Edited by Renate Schmidt. *Proceedings of Conference on Automated Deduction (CADE).* Montreal, Canada: Springer. 151–156.

Brehmer, S. 2013. The Multicore Association Communications API. *The Multicore Association Communications API.* The Multicore Association, 3: 1–5.

Clarke, E., O. Grumberg, and D. Peled. 1999. *Model Checking.* Cambridge, MA: MIT Press.

de Moura, L., and N. Bjørner. 2008. Z3: An efficient SMT solver. Vol. 4963/2008, Chapter 24 in *Tools and Algorithms for the Construction and Analysis of Systems*, 337–340. Berlin: Springer. doi:10.1007/978-3-540-78800-3_{2}{4}.

Dutertre, B., and L. Moura. 2006. A fast linear-arithmetic solver for DPLL(T). Vol. 4144, in *Computer Aided Verification*, edited by Thomas Ball and Robert Jones, 81–94. Berlin/Heidelberg: Springer.

Elwakil, M., and Z. Yang. 2010. *Debugging Support Tool for MCAPI Applications.* Workshop on Parallel and Distributed Systems: Testing, Analysis, and Debugging (PADTAD - VIII). Trento, Italy: ACM.

Foundation, Eclipse. 2013. Eclipse CDT. http://www.eclipse.org/cdt/.

Gait, J. 1986. A probe effect in concurrent programs. *Software: Practice and Experience* (John Wiley &; Sons, Inc.) 16 (3): 225–233. doi:10.1002/spe.4380160304.

Lee, D., M. Said, S. Narayanasamy, Z. Yang, and C. Pereira. 2009. Offline symbolic analysis for multi-processor execution replay. *Proceedings of the 42nd Annual IEEE/ACM International Symposium on Microarchitecture.* New York, NY, USA: ACM. 564–575.

Lee, J., D. Padua, and S. Midkiff. 1999. Basic compiler algorithms for parallel programs. *SIGPLAN Not.* (ACM) 34 (8): 1–12.

Marques-Silva, J. 2008. Practical Applications of Boolean Satisfiability. *Workshop on Discrete Event Systems (WODES'08).* IEEE Press.

Savage, S., M. Burrows, G. Nelson, P. Sobalvarro, and T. Anderson. 1997. Eraser: A dynamic data race detector for multithreaded programs. *ACM Transactions on Computer Systems* 15: 391–411. doi:10.1145/265924.265927.

Sen, K., and G. Agha. 2005. Detecting errors in multithreaded programs by generalized predictive analysis of executions. *Proceedings of 7th IFIP International Conference on Formal Methods for Open Object-Based Distributed Systems (FMOODS'05).* LNCS. Springer. 211–226.

Sharma, S., G. Gopalakrishnan, E. Mercer, and J. Holt. 2009. MCC: A runtime verification tool for MCAPI user applications. *FMCAD.* 41–44.

Wang, C., S. Kundu, M. Ganai, and A. Gupta. 2009. Symbolic predictive analysis for concurrent programs. *FM'09: Proceedings of the 2nd World Congress on Formal Methods.* Berlin, Heidelberg: Springer-Verlag. 256–272. doi:10.1007/978-3-642-05089-3_17.

11 Status of Agile Practices in the Software Industry in 2019

Ashish Agrawal and Anju Khandelwal
SRMS College of Engineering & Technology

Jitendra Singh
Wachemo University

CONTENTS

11.1 INTRODUCTION

The methodology of old software development processes such as the waterfall model (developed by Royce in 1970s) and RAD (rapid application development) (developed by Martin in 1980s) was focused on the sequential working approach; however, with the development of spiral model (Bohem 1986), the development of iterative and incremental models had begun. Later, the Scrum method (developed by Schwaberin 1995) was developed which focused more on iterative and incremental software development (Koç and Aydos 2017). Scrum method then became a part of Agile software development (ASD) whose manifesto was published by 17 software engineers in Utah (United States) in 2001.

ASD is a group of various software development methods that works in an iterative and incremental manner with cross-functional, self-organizing teams. Agile software development methodology (SDM) favors continuous improvement, better communication, changing requirements and an adaptive approach (Ćirić and Gračanin 2017). It values interactions over processes and tools, working software over documentation, customer collaboration over contract negotiation and change overplan (Hohl et al. 2018). In addition to these good features, Agile methods have been criticized for being extreme and inefficient for large organizations, for showing lack of upfront planning, being unsuitable for sequential projects, among others.

ASD is a collection of different programming improvement strategies that work in an iterative and steady manner with cross-practical, self-sorting out-groups. Agile SDM favors nonstop improvement, better correspondence, changing prerequisites and versatile methodology, among others. The customer is always given high priority. During the Agile development, customers are usually closer and more involved with the process making it easier to provide constant feedback (Curcio et al. 2019). It emphasizes on associations over procedures and instruments, working programming over documentation (Ozkan 2019), client coordinated effort over agreement exchange and change over arrangement. In addition to these features, Agile techniques are scrutinized for being extraordinary and wasteful for large associations, for demonstrating the absence of forthright arrangement, being unacceptable for successive tasks, and so forth. The Agile methodology additionally offers services to firms to develop the correct product, which makes them aggressive within the marketplace (Mathur and Satapathy 2019).

11.2 AGILE PRACTICES IN THE SOFTWARE INDUSTRY

ASD is an umbrella term and has many methods and techniques associated with it. Some major Agile methods include Scrum, extreme programming, Kanban and feature driven development pair programming, among others. Each method has its own capabilities and qualities and can be used in a specified area. One of the most used Agile methods is "Scrum", which has been adopted by the entire software industry (Srivastava et al. 2017; Ereiz and Mušiš 2019).

The major reasons behind adopting ASD are fast delivery of software products, the capability to include changing requirements, improved quality of software and increased productivity (Ahuja and Priyadarshi 2015). On-time product delivery is always required to improve customer satisfaction and increase business value, and in ASD, the role and satisfaction of customers have always been the most important aspect. Moreover, ASD leads to reduced project costs (Alzoubi et al. 2018). When product development teams use these methods in various projects, they realize the suitability of an Agile method. The principles and values of the Agile Manifesto are not limited to software development (Conforto et al. 2014) but are also used in cost estimation, product manufacturing, decision making, effort estimation, team collaboration, etc. The major challenges for the software industry to apply the concepts of ASD include the difference between the cultural values of an organization and Agile Values, lack of financial support, resistance to change and upgradation, lack of Agile skills and training, lack of communication between teams, etc. According to Alahyari et al. (2017), the most common barriers for delivery process with respect to time are the unclear definition of delivery, late scope changes and dependencies on other development teams. Many classical Agile projects have difficulties with scalability (Kettunen and Laanti 2017) and dealing with quality factors, such as security and safety. The developers would build some working software and then try to respond to change formaking it safe, secure and scalable, but would find that no simple refactoring would make it safe, secure and scalable (Boehm et al. 2019).

According to the 13th version of the "State of Agile" survey report (CollabNet Version one 2019), the following figures show the top five techniques and various considerations required for implementing ASD in the software industry. These results were based on survey responses collected from 1,319 respondents (CollabNet Version one 2019). Figure 11.1 shows the top five agile techniques selected by the respondents and Figure 11.2 shows the major tips and considerations that are helpful for the success of an Agile project.

The basic idea behind ASD is to develop the product iteratively and incrementally so the whole development is divided into short periods called "sprints" (majorly used in the Scrum method of Agile). Each sprint can last from a couple of days to aweek. Everyday team members meet each other for 10–15 minutes and discuss the module they are working upon, and these short meetings are known as daily stand up meetings. These daily stand-up meetings reduce the full-fledged meeting room settlement cost and increase transparency and awareness among team members. It is very important to have a good sprint planning and implementation.

At the end of sprints and iteration, a retrospective is held to discuss the scope of improvement in the next iteration. Each retrospective focuses on what has happened as well as what is left, and identifies the reasons and solutions for the problems.

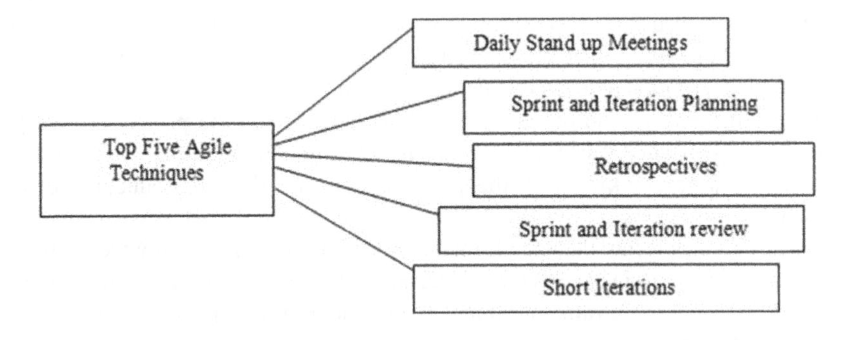

FIGURE 11.1 Major Agile techniques.

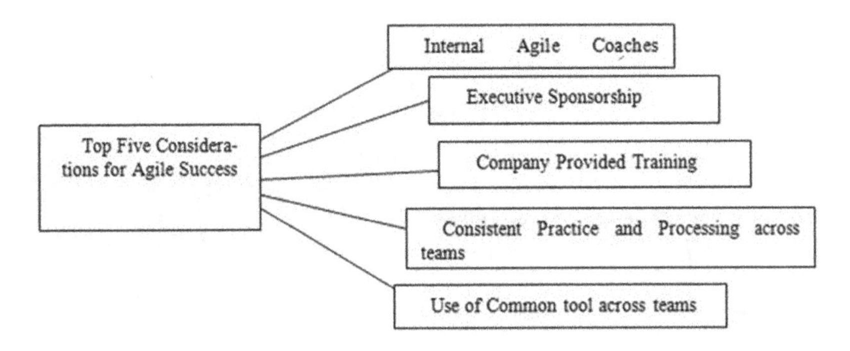

FIGURE 11.2 Tips for Agile success.

11.3 REAL-LIFE EXPERIENCE OF AGILE TEAMS

While working with ASD, the team has the following core pillars that need to be kept in mind:

- Continuous improvement.
- Reducing waste.
- Delivering values that are more and more customer satisfaction oriented.
- It is back to the classroom approach. If you write on it here, people in the United States can see what you are writing and collaborate in real time.

The basic goal would be met at each stage, called a sprint, which can last as little as a week. Each version is deployed so that clients can see what they are getting at an earlier stage and need not wait for the end result. The process is expedited as everyone in the team works closely together. The benefits are also from what you don't have to do than what you have to. The methodology eliminates a lot of work. Things such as management communication, PowerPoint presentations and all of that are not required because everyone knows everything.

The existing challenges faced by the team in the implementation of Agile methodology is day-to-day changing requirements and constant rework for a better understanding of customer's requirements. Mid-level people managers need to go back to the classroom to brush up their skills. It's a bold ambition and it's going to be very difficult as no one has pioneered that road before. But in the digital pod, no one has a cubicle. Every wall is a board that can be written on.

In fact, in many IT industries, millions of dollars from the project are wasted as they don't guarantee success and the waterfall model is often blamed for it. The Agile method aims at taking away the uncertainty around the end product by deploying software in stages.

Figure 11.3 represents the search results for the topic ASD worldwide in all the categories for the past 5 years on Google trends. A value of 100 indicates the superlative for the ASD here. A value of 50 means that the term is half as popular. A value of 0 means that there was not enough data available for the term. Here, a maximum of 100 points is related to the region St. Helena between the periods of January 21 and 27, 2018.

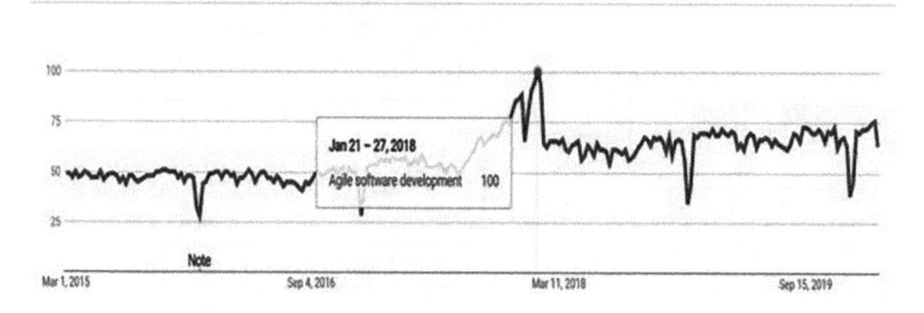

FIGURE 11.3 ASD worldwide in all categories.

Figure 11.4 represents the searching results for the topic ASD worldwide in the category of education for the past 5 years. Here, the values of various points in Figure 11.4 represent the researcher's interest in exploring relative to the highest point on the chart for a given area and time. A value of 100 indicates the superlative for the ASD. A value of 0 means that there was not enough data available for this term. A maximum of 100 points is related to the region St. Helena between the period of September 8–14, 2019. Individually, the peak period of ASD (100 points) was during August 23–29, 2015. Also, a maximum of 45 points is related to the region St. Helena between the period of December 22–28, 2019. This variation is due to the development of the software industry.

Figure 11.5 represents the searching results of ASD worldwide in the category of jobs for the past 5 years. Here, a maximum of 100 points is related to the region Netherlands between the periods of February 9–15, 2020. Individually, in this category, during December 1–7, 2019, the value is 60. This variation is due to the development of the software industry. Globally, India is in the fourth place with a score of 38. Individually, in India, the scoregoes up to 100 during May 3–9, 2015. Currently, during February 9–15, the score is 35.

In Figure 11.6, a maximum of 100 points is related to the region of Karnataka in India during January 19–25, 2020. In the recent period of February 9–15, 2020, the score is 63. This variation is due to the development of the software industry. During March 5–11, 2017, the score was 96, almost the peak time measure.

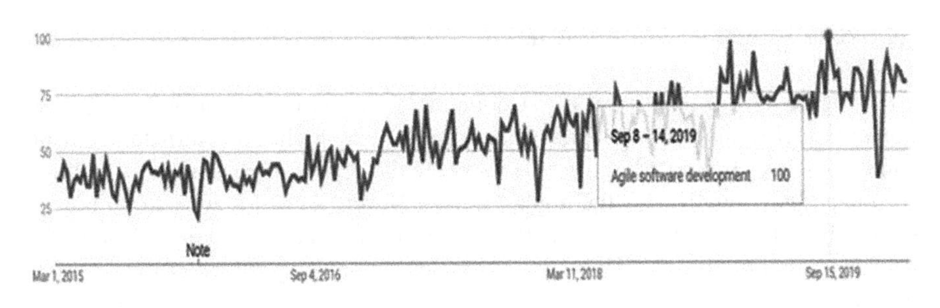

FIGURE 11.4 ASD worldwide in education.

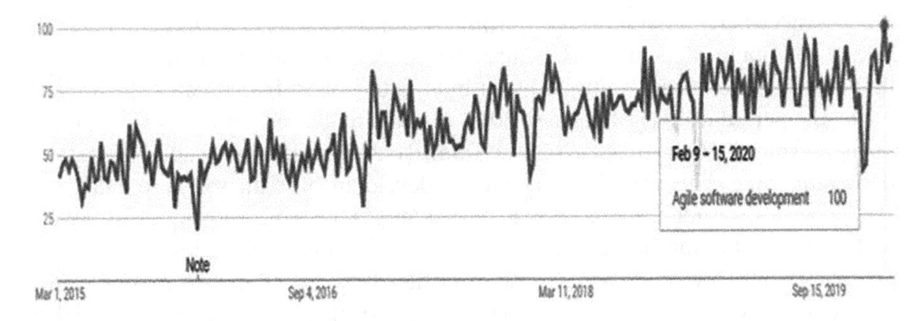

FIGURE 11.5 ASD worldwide in jobs.

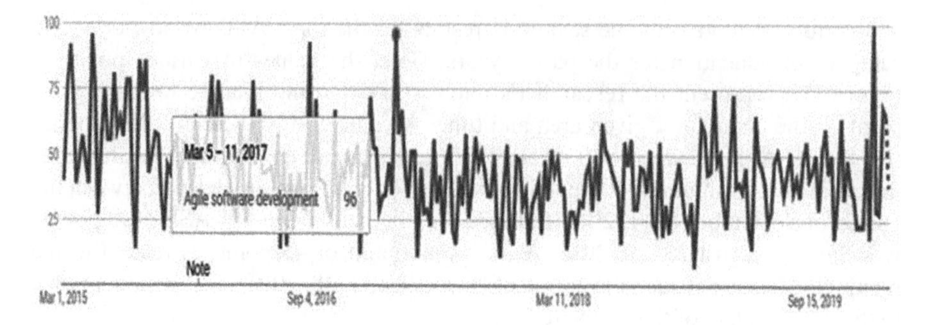

FIGURE 11.6 ASD in India in jobs.

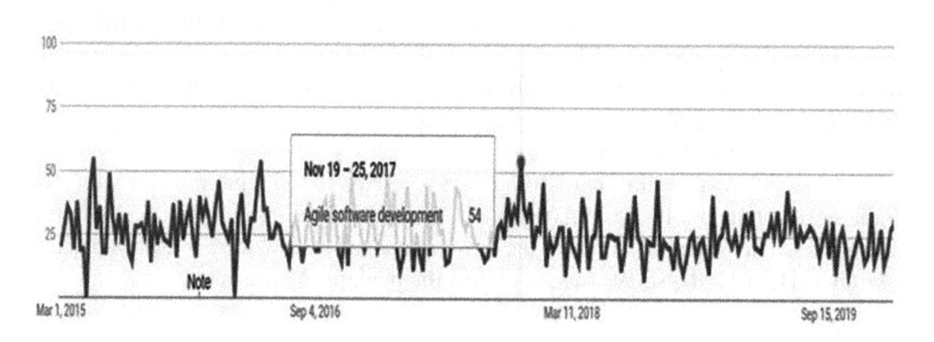

FIGURE 11.7 ASD in India in education.

In Figure 11.7, a maximum of 54 points is for India in the category of education during November 19–25, 2017. State-wise, it is the highest in Karnataka during August 20–26, 2017, at 100 points. Haryana is the second highest state in India at 59 points. Individually, it is the highest during November 8–14, 2015, at 63 points. Currently, during February 2–8, 2020, it is at 30 in Haryana and 25 in Karnataka for the same duration of time. This variation is due to the development of the software industry.

In Figure 11.8, a maximum of 100 points is for India in all the categories during June 14–20, 2015. Also, in Figure 11.6, another maximum point is 93 during November 19–25, 2017. Here, between the above period, ASD has grown very fast almost everywhere. State-wise, it is the highest in Karnataka during December 6–12, 2015 at 100 points, and during December 10–16, 2017, it is at peak point (99 points). Tamil Nadu is the second-highest state in India at 69 points. Individually, it is highest during the period of May 31–June 6, 2015, at 100 points. Currently, it is at 82 points during February 26–March 1, 2020 in Tamil Nadu and at 64 points in Karnataka at the same time. This variation is due to the development of the software industry.

The waterfall model method is also known as the linear sequential life cycle model. Unlike the waterfall model, in this model, development and testing activities are concurrent. DevOps is an exercise to bring development and operations teams together (Doukoure and Mnkandla 2018), whereas Agile is an iterative approach that focuses on collaboration, customer feedback and small rapid releases. In Figures 11.9

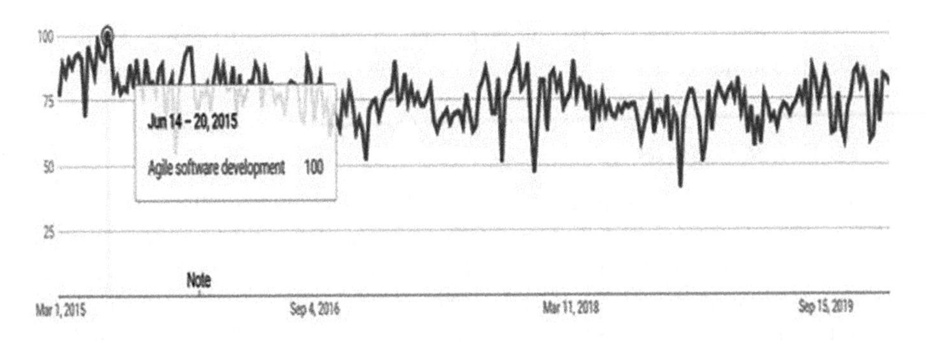

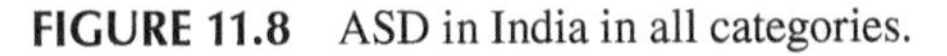

FIGURE 11.8 ASD in India in all categories.

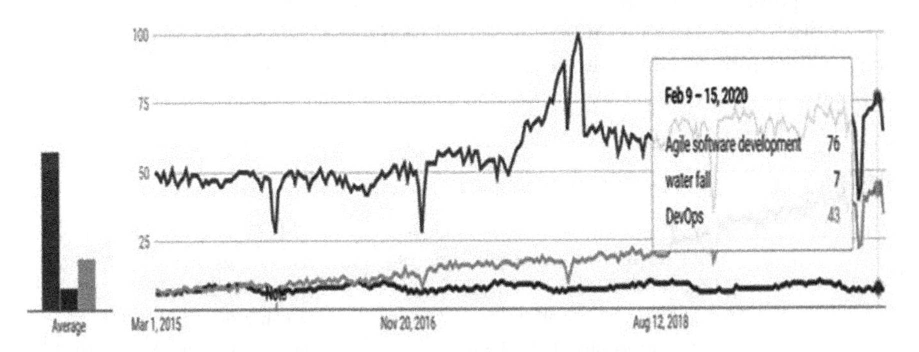

FIGURE 11.9 Comparison of ASD, WF and DevOps worldwide in all categories.

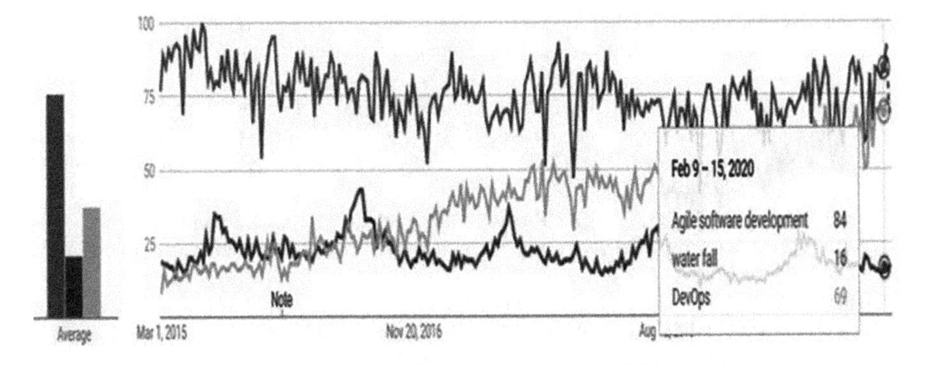

FIGURE 11.10 Comparison of ASD, WF and DevOps in India in all categories.

and 11.10, a comparison of ASD, waterfall and DevOps is shown for both worldwide as well as India in all categories.

Figures 11.9 and 11.10 show the ASD, waterfall and DevOps (scores (76, 7, 43) worldwide and (84, 16, 69) in India during February 9–15, 2020, in all categories. In both the above graphs, waterfall has the least measure 7 and 16, which represents that in the current environment this model is not much in use. Moreover, we have seen that DevOps is an upcoming new approach that is better than ASD. In India, the Agile approach is continuously decreasing in use from 2015 onward whereas DevOps is continuously increasing during the same period.

11.4 CONCLUSION

ASD has changed the software industry in a well-defined manner. The customers are not just the stakeholders now but they are part of the whole software development. The major parameter of the success of ASD is its approach to changing requirements. Customers are now more satisfied and developers are getting high returns on investment. In conclusion, we can say that Scrum is the most important Agile method that has been used by most of IT organizations. ASD has changed software development and has spanned almost the entire global IT industry. At present, ASD is facing huge competition with DevOps, which is an amalgamation of development and operations. In the future, we might see a technological shift from ASD to DevOps in the IT industry.

REFERENCES

Ahuja L. and R. Priyadarshi 2015. Agile approaches towards Global Software Engineering. *4th International Conference on Reliability, Infocom Technologies and Optimization (ICRITO) (Trends and Future Directions)*, Noida, IEEE Xplore, DOI: 10.1109/ICRITO.2015.7359249: 1–6. https://ieeexplore.ieee.org/document/7359249.

Alahyari H., R. B. Svensson and T. Gorschek 2017. A Study of Value in Agile Software Development Organizations. *The Journal of Systems and Software* 125: 271–288. DOI:10.5555/3063155.3063214.

Alzoubi Y. I., A.Q. Gill, B. Moulton 2018. A measurement model to analyze the effect of agile enterprise architecture on geographically distributed agile development. *Journal of Software Engineering Research and Development* 6, 4. DOI: 10.1186/s40411-018-0048-2. https://link.springer.com/article/10.1186/s40411-018-0048-2.

Boehm B., D. Rosenberg and N. Siegel 2019. Critical Quality Factors for Rapid, Scalable, Agile Development. *IEEE 19th International Conference on Software Quality, Reliability and Security Companion (QRS-C)*. 978-1-7281-3925-8/19: 514–515. https://ieeexplore.ieee.org/document/8859476.

Ćirić D., D. Gračanin 2017. Agile Project Management beyond Software Industry. *XVII International Scientific Conference on Industrial Systems*. http://www.iim.ftn.uns.ac.rs/is17:332–337.

CollabNet Version one 2019. 13th Annual State of Agile Report. http://www.stateofagile.com/?_ga=2.88255835.1987726296.1584501644-642308080.1584501644#ufh-i-521251909-13th-annual-state-of-agile-report/473508.

Conforto C. E., F. Salum, D. C. Amaral, S. L. Silva, L. F. M. Almeida 2014. Can Agile Project Management Be Adopted by Industries Other than Software Development? *Project Management Journal*. 21–34. DOI: 10.1002/pmj. https://journals.sagepub.com/doi/10.1002/pmj.21410.

Curcio K., R. Santana and S. Reinehr and A. Malucelli 2019. Usability in Agile Software Development: A Tertiary Study. *Computer Standards & Interfaces* 64: 61–77.

Doukoure K. A. G., E. Mnkandla 2018. Facilitating the Management of Agile and Devops Activities: Implementation of a Data Consolidator. *International Conference on Advances in Big Data, Computing and Data Communication Systems, Durban, IEEE Xplore*, DOI:10.1109/ICABCD.2018.8465451: 1–6.

Ereiz Z., D. Mušić 2019. Scrum Without a Scrum Master. *IEEE International Conference on Computer Science and Educational Informatization (CSEI)*, Kunming, China, DOI:10.1109/CSEI47661.2019.8938877: 325–328. https://ieeexplore.ieee.org/document/8938877.

Hohl, P., J. Klünder, A. V. Bennekum et al. 2018. Back to the Future: Origins and Directions of the "Agile Manifesto" – Views of the Originators. *Journal of Software Engineering Research and Development* 6, 15. DOI: 10.1186/s40411-018-0059-z.

Kettunen P., M. Laanti 2017. Future Software Organizations – Agile Goals and Roles. *European Journal of Future Research* 5, SpringerOpen. DOI:10.1007/s40309-017-0123-7. https://link.springer.com/article/10.1007/s40309-017-0123-7.

Koç G., M. Aydos 2017. Trustworthy Scrum: Development of Secure Software with Scrum. *International Conference on Computer Science and Engineering (UBMK), IEEE*, DOI:10.1109/UBMK.2017.8093383: 244–249.

Mathur B. and S. M. Satapathy 2019. An Analytical Comparison of Mobile Application Development Using Agile Methodologies. *Proceedings of the Third International Conference on Trends in Electronics and Informatics*, IEEE Xplore Part Number: CFP19J32-ART: 1147–1152.

Ozkan N. 2019. Imperfections Underlying the Manifesto for Agile Software Development. *1st International Informatics and Software Engineering Conference (UBMYK)*, Ankara, Turkey, IEEE Xplore, DOI: 10.1109/UBMYK48245.2019.8965504: 1–6. https://ieeexplore.ieee.org/document/8965504.

Srivastava A., S. Bhardwaj and S. Saraswat 2017. SCRUM model for agile methodology. *International Conference on Computing, Communication and Automation (ICCCA)*, Greater Noida, IEEE, DOI: 10.1109/CCAA.2017.8229928: 864–869. https://ieeexplore.ieee.org/document/8229928?section=abstract.

12 Agile Methodologies
A Performance Analysis to Enhance Software Quality

Neha Saini and Indu Chhabra
Panjab University

CONTENTS

12.1 INTRODUCTION

Agile software development methodologies are gaining popularity among various software development organizations as they are customer-oriented and productive. The limitations of the traditional approach gave rise to the need for a methodology that can replace the flaws of the traditional approach with the strengths of a flexible approach. Agile practices are adaptable to changes and involve customers in the development process. The inflexible traditional approaches were unable to compete with these lightweight software development methods, which could well understand and cope with the customer's demands. Agile software development processes help developers to complete projects in a very short time frame, leading to their popularity. Many studies and surveys are being conducted on agile methods, and software development has entered a new era. As there are diverse and multiple agile methodologies in software engineering, choosing the appropriate one is indeed a challenging quest. The agile processes have started invading the software development industry to provide good quality software in minimal time. In order to achieve success in any software development, it is necessary that an appropriate methodology should be adopted, as choice of wrong methodology may increase the failure rate of the software to be developed or completion of the software ahead of budget and the timeline (Khan and Beg 2013). Since agile methodologies are more flexible as compared to traditional ones this chapter aims to analyze and compare the varied agile methodologies that help in improving the software quality by choosing the appropriate one. It also aims at analyzing the role of artificial intelligence in solving the above stated problem.

12.2 PROBLEM DEFINITION

In this study, the problem statement revolves around developing a comprehensive selection paradigm to determine the suitability of agile approaches for a given software project. To achieve the desired result, at the outset, the characteristics of the software such as project type, duration, cost, complexity and other parameters will be analyzed, and then a selection paradigm will be developed to determine the suitability of an agile approach for the given project. This study aims to provide guidance to the software developers regarding how to select agile practices and tune the selected practices to the given project environment.

12.3 RESEARCH QUESTIONS

The following research questions have been endeavored to be answered in this study:

- What are the characteristics of the software based on user's requirements?
- What are the different agile methodologies that can be mapped with the identified characteristics?
- What should be an appropriate methodology to be adopted by the software developer in a given project?
- How the responses to the above questions will amalgamate in the form of an extensive selection paradigm?

12.4 AGILE METHODOLOGY

The agile methodology is a recent development framework, which is used as a solution to the limitations of the traditional approaches in software engineering (Dingsoeyr et al. 2019). In developing good quality software, various other problems emerged. Some of these problems include:

4.1 Changing needs of the customer: As user needs are dynamic, they do not have a clear vision about the requirement specification during the early stages. Only when an application reaches them, they realize their true needs from the software.

4.2 Customer involvement: As software projects do not involve customers beyond the requirement specification stage, it increases the chances of project failure.

4.3 Deadlines and budgets: Tight deadline and low budget along with the correct high-end software leads to competitions between developers.

4.4 Miscommunication: Due to the gap between customer expression and understanding by the developers of the software requirement specifications, a lot of misunderstanding of the customers' need is created.

12.5 AGILE PRINCIPLES

The concept of agile development was proposed in 2001 and the Agile Manifesto was published wherein the 12 principles were documented. The primary principle was to give the utmost importance to customer satisfaction by quickly delivering the software and accepting the changing needs of the user at any stage of the development process. The other principles involve reducing the delivery time of the software, coordination among the users and the software developing team during project development, as well as to trust, support and provide a congenial environment to the project team in developing the software. (Hoda et al. 2017) The following principles have been recognized as per Agile Manifesto, 2001 (Beedle et al. 2002):

- Quick delivery of software to the customer.
- Adapt to the changing needs at later stages.
- Delivering working software frequently in a shorter timescale ranging from weeks to months.
- Collaborative work of teams during the entire project.
- Work in motivational environment for early completion of tasks.
- In-person interactions for sharing and discussing information.
- Workability of the software as a means to measure the progress.
- Promoting sustainable development.
- Constant consideration to technical expertise and good design augments agility.
- Simplicity is pertinent for maximum and optimum results.
- Self-organization of teams is the key to best architectures and designs.
- Ability of the team to adjust behavior at regular intervals to make it more effective.

12.6 FRAMEWORK FOR AGILE PRACTICES

Agile methodologies constitute a group of approaches which are more flexible in nature, that is, they are adaptable to the changing needs of the user, even at later stages, and involves user in the development process. Some of the existing agile practices are

6.1 Scrum Methodology: This method contains many features of the iterative, incremental and traditional methodologies. The primary objective in the scrum method is to use the concept of sprint. A sprint consists of a duration of 30 days along with a group of specified objectives. The process of scrum begins with a planning stage. In this phase, a backlog list is created that provides the details pertaining to the functionality of various releases (Magana et al. 2018).

6.2 **Extreme Programming (XP)**: Another important and commonly used practice is extreme programming. It involves a group of different agile approaches. The process of XP initiates with the planning phase in which the feasibility of software from different perspectives, the hard work needed and the timeline for the first release are examined and evaluated. After this, the other important characteristics of the software are defined in the form of stories (Moniruzzaman and Hossain 2013). The rationale of XP method is to keep the blueprint straightforward to the optimum.

6.3 **Feature-Driven Development (FDD)**: The FDD methodology is also an important agile practice. This approach differs from other practices in a sense that it emphasizes on planning and open design. The primary focus in the development process is on the characteristics of the system. Therefore, it requires various other supporting methodologies for developing a software

6.4 **Dynamic System Development Methodology (DSDM)**: It was developed in 1990s in the United Kingdom. This approach differs from other approaches in the sense that the main idea is reversed in this approach compared to other agile approaches. In DSDM, the resources and schedules are fixed, and then the functional requirements are adjusted according to the fixed resources.

6.5 **Adaptive Software Development Methodology (ASD)**: It comprises three phases, speculate, collaborate and learn (Moniruzzaman and Hossain 2013).In the first phase, whatever we need to achieve after completing each pass is defined. The significance of team work by sharing information between the development team is emphasized during the second phase. After this, the third phase is executed after every pass so as to enhance the skills of the developer.

12.7 BENEFITS AND COMPARISON OF AGILE TECHNOLOGIES

Because of the existence of multiple agile practices, there arises a need to analyze which methodology shall be suitable for the given requirements of the user (Matkovic and Tumbas 2010). The use of agile practices for software development helps the team to accept the changes given by the customers at each stage, as well as helps the developers to reduce the risk during the initial stage of project development. The developers get a chance to learn from mistakes at each stage of development, which reduces the time lag and enhances the quality of the developed software (Petersen and Wohlin 2009). The comparison of agile methodologies with their respective features and identified weaknesses is summarized in Table 12.1 (Fernando et al. 2013).

12.8 PRECINCTS OF USING AGILE METHODOLOGY

Agile methodologies require considerable customer involvement while developing a software, so it may pose difficulties for customers who do not have enough

TABLE 12.1
Comparison of Agile Technologies

Agile Method	Characteristics	Weakness
Scrum	Small, independent and self-organizing development teams, 30-day sprint	No details are provided with the system and acceptance testing during the release sprint of 30 days
XP	User-oriented, small teams. Effort needed and release timeline are examined and evaluated	The overall view is not focused and emphasized
FDD	The emphasis is on planning and upfront design	Not a complete methodology as it emphasizes only on design and requires other supporting approaches
DSDM	Resources and schedules are fixed and then the functional requirements are adjusted accordingly; quick application development	Not enough clarity on the actual use of methodology as only members have access to white papers related to its use
ASDM	Enhances and improves the skills of the developer and iterative development	Focus is more on the concepts rather than practical implementation

time for such participation. The situation may worsen if the development team is not located at the same physical location as agile practices support close working associations for easy handling of the projects (Shankarmani et al. 2012). Moreover, as agile approaches support time-bound delivery, there is a possibility that few targets may not get completed within the prescribed time limit (Hajjdiab and Taleb 2011). The repetitive behavior of agile development may influence the system's behavior in general, due to lack of focus on understanding of the developed system as a whole at the initial phases of software development. It becomes more relevant in complex systems with large implementations or with software that involve high level of integration. In agile approaches, the focus is only on the code, which may lead to memory loss, as documentation done is not substantial (Hneif and Hockow 2009). A software designed using agile methodologies does not provide a reusable version as they build problem-specific systems rather than a general system, leading to decrease in the quality and performance of the software.

12.9 METHODOLOGY

To deliver a project within budget, within deadlines and according to the needs of the user, the selection of an appropriate methodology plays an important role in the development of the project. However, because of the existence of multiple agile practices, there arises a need to anatomize as to which methodology shall be suitable for the given user requirements. To achieve this goal, the present study intends to propose a selection paradigm to resolve the above-stated problem. The paradigm will help the developers to choose the appropriate agile methodology while developing a project. This will contribute toward the efficiency in the arena of software development methods. The study has been carried out as follows:

- To determine the appropriate methodology for the software, first, there is a need to understand the project itself, which further requires the need to classify the project according to the characteristics based on the user requirements (Jorgensen 2019). At this stage a high-level assessment of project facets is conducted to determine the applicability of agile methods to software projects. The facets so considered are the following (Yusof et al. 2011):
  - Size
  - Complexity
  - Criticality
  - Stability
  - Development timeframe
  - Risk factor

The mapping of above parameters has been done with two other important parameters before considering the final choice:

- Involvement and priority of the customer
- Organization-specific factors (policies and procedures followed and structure of the organization)

A rating method has been employed to provide a more systematic and flexible approach to methodology selection. A score has been assigned to all the above parameters ranging from 0 to 5, where low scoring means that the agile methodologies are suitable while a high scoring leads the development preference towards traditional methodologies. This step gives an indication to the probable preference of the method, i.e., whether traditional methodologies are suitable or agile are suitable.

- In this step, the existing agile methodologies and their phases and factors responsible for the agile methodology selection are ascertained (Nerur et al. 2005).The goal is to find out the application of suitable agile methods for developing a software and selecting the appropriate one from the existing methods. It is figured out that the project attunes to one or more than one agile methods. If the requirements attune to a single agile methodology, then that methodology can be used; otherwise further steps are implemented. The methodology parameters are the means and mechanisms used, the mode scope, adoption, acts, obligations and support for development teams.
- In this step, the organization-specific environmental factors are identified as they may also affect the methodology adoption to a significant extent.
- Associations and mapping among the above are then identified to determine how change in one affects the other. A scoring is also given while mapping and the requirement for choosing the technique that conforms to

the stated problem from among the available methodologies (Venera et al. 2011). Then a probable solution is devised in the form of guiding framework for developers, which will help to select the optimum agile method.

12.10 ROLE OF AI TECHNIQUES

The role of artificial intelligence (AI) techniques in solving the above problems has also been analyzed. The machine learning techniques also play a very important role in the software development as:

- **Coding Assistants Implemented Using Machine Learning**: With smart coding assistants, developers can quickly receive feedback on to the basis of codebase, saving a lot of time of the developer which is usually spent on debugging and reading documentation, etc.
- **Automatic Coding Refactoring**: With machine learning and identifying potential areas for refactoring, it is easy to analyze the code and its optimization for performance.
- **Prioritization**: An AI model helps the teams to identify the methods of minimizing risk and maximizing impact.
- **Predicting Correct Estimates**: The machine learning model can also be proved helpful in predicting the precise estimates and budgets.
- **Error Handling**: The smart coding assistants implemented using machine learning are beneficial in identifying errors.
- **AI in Planning**: Machine learning can create various combinations and permutations of a situation similar to the human brain while planning for a project.
- **Risk Estimation**: The machine learning model can also be helpful in predicting the risks earlier.

12.11 CONCLUSION

Lack of proper project development experience, technical incompetence of software developers, dearth of proper design and inadequate methods of implementation have changed the software development scenario in the contemporary era of technology. The iterative nature and user involvement in agile methodology is responsible for its increased use and acceptance in software development industry. Hence, novel techniques for software development are the need of the hour to regulate the complications of the massive software systems to enable the delivery of the software well on time and within budget. Though agility brings quality, it has its own shortcomings. A careful decision has to be made for each project as in some cases traditional methods would score over agile methodology. Negligence to choose an apt agile method may also lead to software failure or completion of software ahead of budget and the timeline.

REFERENCES

Beedle, M., Bennekum, A.V., Cockburn, A., Cunningham, W., Fowler, M., and Highsmith, J. 2002. Agile manifesto. http://agilemanifesto.org/principles.html. (accessed December 12, 2019).

Dingsoeyr, T., Falessi, D., and Power, K. 2019. Agile development at scale: The next frontier. *Software IEEE* 36:30–38.

Fernando, W.K.S.D., Wijayarathne, D.G.S.M., Fernando, J.S.D., Mendis, M.P.L., and Menwadu, C.D. 2013. Emergence of agile software development methodologies: A Sri Lankan software research and design outlook. *IJSTR* 2:103–108.

Hajjdiab, H., and Taleb, A.I.S. 2011. Adopting agile software development: Issues and challenges. *IJMVSC* 2:1–10.

Hneif, M., and Hockow, S. 2009. Review of agile methodologies in software development. *International Journal of Research in Applied Sciences* 1:1–8.

Hoda, R., Salleh, N., Grundy, J., and Tee, H. 2017. Systematic literature reviews in agile software development: A tertiary study. *Information and Software Technology* 85, no. 1 (May):60–70. https://www.sciencedirect.com/science/article/abs/pii/S0950584917300538.

Jorgensen, M. 2019. Relationships between project size, agile practices, and successful software development: Results and analysis. *IEEE Software* 36, no. 2:39–43. https://doi.org/10.1109/MS.2018.2884863.

Khan, P.M., and Beg, M.M. 2013. Extended decision support matrix for selection of SDLC-models on traditional and agile software development projects. *International Journal of Software Engineering Research and Practices* 3, no. 1 (April):1–8. http://citeseerx.ist.psu.edu/viewdoc/download?doi=10.1.1.676.4737&rep=rep1&type=pdf.

Magana, A., Seah, Y.Y., and Thomas, P. 2018. Fostering cooperative learning with scrum in a semi-capstone systems analysis and design course. *Journal of Information Systems Education* 29, no. 2 (March): 75–92. https://jise.org/Volume29/n2/JISEv29n2p75.pdf.

Matkovic, P., and Tumbas, P. 2010. A comparative overview of the evolution of software development models. *Journal of Industrial Engineering and Management* 1, no. 4:163–172. https://www.academia.edu/3376905/A_Comparative_Overview_of_the_Evolution_of_Software_Development_Models.

Moniruzzaman, A.B.M., and Hossain, S. 2013. Comparative study on agile software development methodologies. *Global Journal of Computer Science and Technology* 13, no. 7 (July):5–18. https://www.researchgate.net/publication/249011841_Comparative_Study_on_Agile_software_development_methodologies.

Nerur, S., Mahapatra, R., and Mangalraj, G. 2005. Challenges of migrating to agile methodologies. *Communications of the ACM* 48:72–78.

Petersen, K., and Wohlin, C. 2009. A comparison of issues and advantages in agile and incremental development between state of the art and an industrial case. *Journal of Systems and Software* 82, no. 9 (September):1479–1490. https://www.sciencedirect.com/science/article/pii/S0164121209000855.

Shankarmani, R., Pawar, R., Mantha, S., and Babu, V. 2012. Agile methodology adoption: Benefits and constraints. *International Journal of Computer Applications* 58, no. 15 (November):31–37. https://www.researchgate.net/publication/261017281_Agile_Methodology_Adoption_Benefits_and_Constraints.

Venera, C., Jianu, I., Jianu, I., and Gavrila, A. 2011. Influence factors for the choice of a software development methodology. *Accounting and Management Information Systems* 10:479–494.

Yusof, M., Shukur, Z., and Abdullah, A. 2011. Cuqup: A hybrid approach for selecting suitable information systems development methodology. *Information Technology Journal* 10, no. 5 (March):1031–1037.https://scialert.net/fulltextmobile/?doi=itj.2011.1031.1037.

13 Pretrained Deep Neural Networks for Age Prediction from Iris Biometrics

Ganesh Sable
Maharashtra Institute of Technology

Murtaza Mohiuddin Junaid Farooque
Dhofar University

Minakshi Rajput
Dr. Babasaheb Ambedkar Marathwada University

CONTENTS

13.1 INTRODUCTION

Biometrics is a reliable and accurate system for the identification of a person from biometric data stored in a database of a system. Biometric-based authentication has applications in almost every field, including criminal investigation, parenthood determination, national ID cards, airport security, border control and online banking. The structure of even of two irises with the same genetic genotype (e.g., as in identical twins, or the iris pair by one individual) has an uncorrelated pattern. The iris texture has no genetic penetrance. Therefore, each iris is unique (Muron and Pospisil 2000). The complex pattern of the iris has many distinguished features such as ligaments, furrows, ridges, crypts, freckles and collarette. (Daugman 2001). An integro-differential operator was developed by Daugman for locating inner and outer boundaries of the iris. Further, he adopted a two-dimensional Gabor wavelet transform to extract the features of the iris (Daugman 1994). Wildes published his research work and a patent on iris recognition (Wildes 1997). Other researchers have implemented end-to-end "ResNet50," a deep learning convolution neural network (CNN) used for iris recognition. The proposed model was trained for 100 epochs using an Nvidia Tesla GPU. They could achieve 95.91% accuracy for the IITD database (Minaee and Abdolrashidi 2019).

A deep-learning-based iris segmentation method was used for iris recognition of postmortem iris The proposed method has reached the Equal Error Rate (EER) less than 1% for samples collected up to 10 hours after death. For samples collected up to 369 hours post-mortem, the proposed method has achieved the EER 21.45%. A fast R-CNN model with six layers was constructed to detect the location of an iris in an image. This iris segmentation method could achieve 95.49% of accuracy for CASIA-Iris-Thousand database (Li et al. 2019).

The structure of the iris develops from the childhood to the adult stage and later deteriorates with age. Deterioration of iris is caused by factors such as decrease in pupil dilation capacity, cataract removal surgeries and alteration of iris structure due to glaucoma. These observations may help in predicting the age of a person whose biometric data is saved in a database of the system (Aslam et al. 2009; Fairhurst and Erbilek 2011; Fenker and Bowyer 2012; Glasser and Kasthurirangan 2006; Lanitis 2001; Li et al. 2019; Browning and Orlans 2014).

In this context, we proposed deep learning techniques for the prediction of the age group of a person from iris biometrics. We conducted experimentation on a real-time database. In this work, deep CNN and pretraineddeep networks, such as AlexNet and GoogLeNet, were adopted for experimentation.

13.2 LITERATURE SURVEY

To our knowledge, only two research studies aimed to identify age from the iris structure. One important research work was conducted by Erbilek and his team for identifying age from the human iris structure. In their experimentations, the authors classified the input iris image into one of the three age groups: children (<25 years), youth (25–60 years) and senior citizen (>60 years). They experimented on the BioSecure Multimodal Database for identifying age from the iris. LG Iris Access EOU3000 system was utilized to acquire all images in this database. This database consists of a total of 200 subjects with varying ages (18–73 years). Only five geometrical features are extracted from the iris. Support vector machine (SVM), Multi-Layer Perceptron(MLP), K-Nearest Neighbor(KNN), and multiclassifier approach (based on fusion and negotiation) are implemented for classification. The authors were able to achieve an accuracy of 75% for the multiclassifier-negotiation approach (Erbilek et al. 2013).

The second approach divided 100 subjects into two age groups: 50 subjects in the younger group (22–25 years) and 50 subjects in the older group (>35 years). They extracted 630 textural features from the iris. The experiment was performed by 10-fold cross-validation approach using the random forest algorithm (Weka using 300 trees). The proposed algorithm achieved an accuracy of 64.68% (Sgroi et al. 2013).

On the basis of the above study, the results mentioned in both approaches seem to be realistic and acceptable. The extraction of 630 features makes a system slower and computationally expensive in the second approach. The main limitation of both studies is that they have not considered kids in their experimentation. In our previous study on true age prediction from the iris image, we achieved good performance for deep CNN and AlexNet. Results for CNN showed that, though our trained CNN model did not reveal the true age for most subjects, predicted age was close to the true age with an error of ±5–7 years (Rajput and Sable 2019).

If iris biometric tries to match the iris of a person after a long period, there may be a possibility of false nonmatch. This is caused by either physiological aging or template aging. Thus, it can be concluded that with growing age, the amount of pupil constriction increases exponentially, resulting in age-related changes in the structure of the iris muscles. This information could be used to reveal the age of a person from the iris. In this context, we tried to develop a system that reveals the age group of a person from his/her iris structure.

13.3 PROBLEM IDENTIFIED

Any organization would be at a high risk if an impostor tries to gain access. The existing biometric systems either accept or reject a person depending upon the match score provided by the system, but the system does not provide any other information such as gender, age, ethnicity, color and height about the impostor. If this type of information is extracted from the existing systems, it will help identify the person who is not enrolled in the system but tries to gain access within a highly secure organization. If the gender is known, searching time is reduced to half.

13.4 COMPARATIVE STUDY OF EXISTING SOLUTIONS

The state-of-the-art methods are compared concerning the following points.

a. Iris scanner, age variation and the number of subjects in databases

   Erbilek and his team experimented on the BioSecure multimodal database consisting of 200 subjects in the age range of 18–73 years. Images in this particular database are captured with the Iris Access EOU3000 system that generates an image with a pixel size of 640×480 (Erbilek et al. 2013). The second approach by Agroi and his team conducted experiments on 100 subjects. Images were captured with scanner LG IrisAccess 4000. The age range of the subjects was 22–25 years and older than 35 years. It means the second study did not have significant age variation compared with the first study (Sgroi et al. 2013).

b. Computation speed

Large feature vector size of 630 makes a system slower and computationally expensive (Sgroi et al. 2013). Database used in both studies have subjects' age above 18 years. This means that they did not include kids in their experimentations.

13.5 PROPOSED SOLUTION

This section describes the processing steps undertaken to predict the age-group from the iris biometrics. Figure 13.1 shows a proposed system block diagram. The system accepts an iris image from the database. This image passes through a step of preprocessing where resizing, reshaping and low-pass filtering and contrast enhancement is done. Preprocessing is necessary to convert an image to a form that is suitable for the next step of neural network training and the final classification. The output of the system is the –age-group of the input iris image. The input image falls under one of the three categories: Kids to young adult, adult and senior citizen.

All the steps in the block diagram are explained next.

13.5.1 Image Acquisition and Database

We used the I Scan 2 scanner to acquire all images in the near-infrared wavelength band (700–900 nm) of the electromagnetic spectrum. All images were acquired in the SAP laboratory of the computer science and engineering department of Dr. BAM University, Aurangabad, Maharashtra, India. I Scan 2 scanner has two 1.3 megapixels camera which simultaneously captures both irises in less than 10 s. It generates an image of 480×480 with 16.7 pixels/mm pixel resolution. The real-time database contains a total of 2,130 images of 213 subjects (125 males and 88 females), including images of 5 left irises and 5 right irises per subject. Figure 13.2 shows age-group-wise male and female sample iris images in the real-time database. Table 13.1 shows the details of a real-time database according to age range.

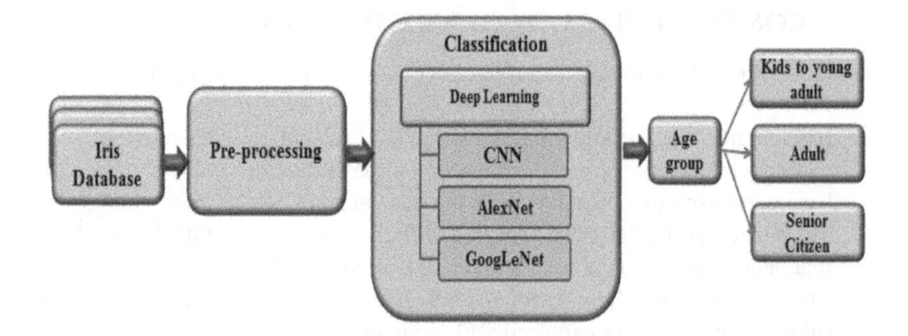

FIGURE 13.1 Block diagram of the proposed system for age-group identification from the iris.

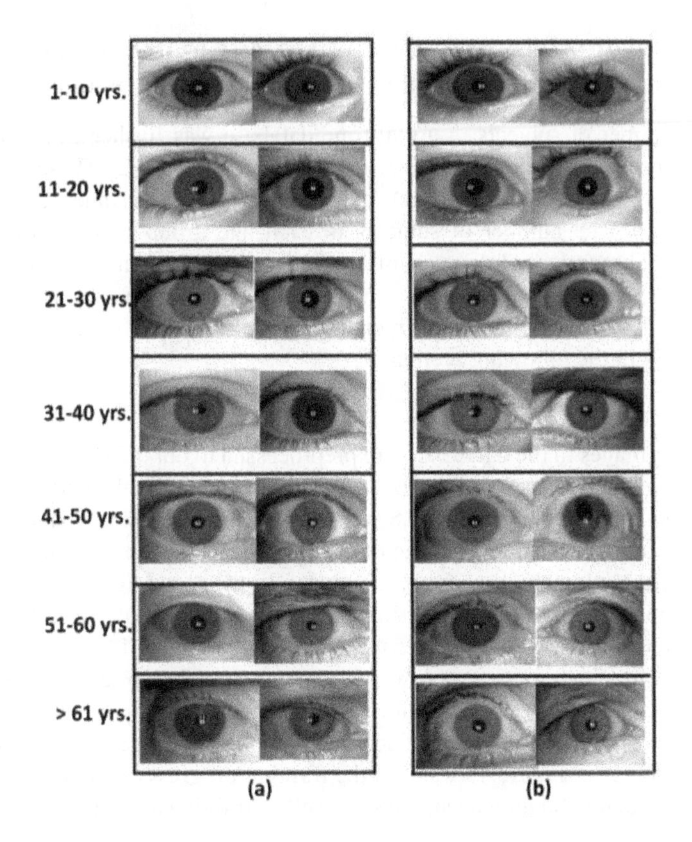

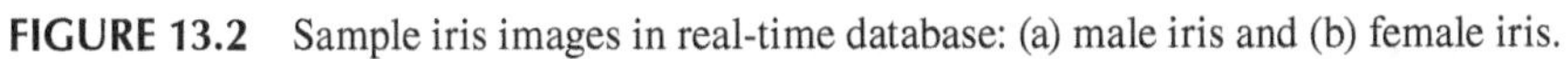

FIGURE 13.2 Sample iris images in real-time database: (a) male iris and (b) female iris.

TABLE 13.1
Details of Real-Time Database

	Number of Subjects		
Age in Years	**Male**	**Female**	**Total**
1–10	12	05	17
11–20	19	14	33
21–30	49	40	89
31–40	13	15	28
41–50	13	04	17
51–60	14	05	19
61–70	02	05	07
71 onward	03	00	03
Total	125	88	213

13.5.2 Experimentation on a Real-time Database to Identify the Age Group from Iris Biometrics Using Deep Learning

The total number of subjects in a real-time database was further categorized into three age groups for experimentation, as shown in Table 13.2. These groups are kids to young adult (age <20 years), adult (age 21–60 years) and senior citizen (age > 60 years). The number of subjects in each age group was 50, 153 and 10, respectively. Three types of deep neural networks are trained on a training database for three age groups. These trained network models are further used to classify the age group of test images. The experimentation details and results of these networks are illustrated in the following sections.

A. CNN

All images in the database were preprocessed to obtain only the iris part of the image and then resized to 200×200. All images in the database were randomly split by a factor of 0.8 so that 80% of the samples from each age group could be used for training the SVM, and the remaining samples were used for testing the age group of images in the test dataset. CNN training options for age group identification are shown in Table 13.3. The time required to train CNN was 16.5 min. As shown in Table 13.4, all test samples from an adult age group were correctly classified except one sample, thus providing good accuracy for the adult group as compared to the other two age groups. Comparison of accuracies for different age groups is shown in Figure 13.3. Figure 13.4 shows the training progress curve for CNN for training and validation accuracies, as well as training and validation loss.

B. AlexNet-based age-group identification

Images in the real-time database are intensity images of size 480×680.

(1) All images in this database were cropped to obtain only the iris part and then resized to $227 \times 227 \times 3$.
(2) The database thus formed was randomly split by a factor of 0.8, so that 80% of the images from each age group form a training database and the remaining 20% from each age group form a test database.
(3) Pretrained AlexNet was loaded.
(4) Features were extracted from fc7 fully connected layers for training and test images. fc7 layer gives a feature vector of size $4{,}096 \times 1$ for each input image.

TABLE 13.2
Experimentation Training and Testing Protocol

Age Groups	Training Images	Testing Images	Total Images
Kids-young adult (<20 years)	200	50	250
Adult (21–60 years)	600	165	765
Senior citizen (>60 years)	40	10	50

TABLE 13.3
CNN Training Options for Age-Group Prediction

Training Options	Values
Training dataset	80% from each class
Test dataset	Remaining 20%
Learning rate	0.003
Epochs	8
Mini batch	128
Validation frequency	30
L2 regularization	1.000 e-04

TABLE 13.4
CNN Accuracies for Different Age Groups

Age Groups	Accuracy (%)
Adult	99.25
Kids-young adult	74.24
Senior citizen	83.33

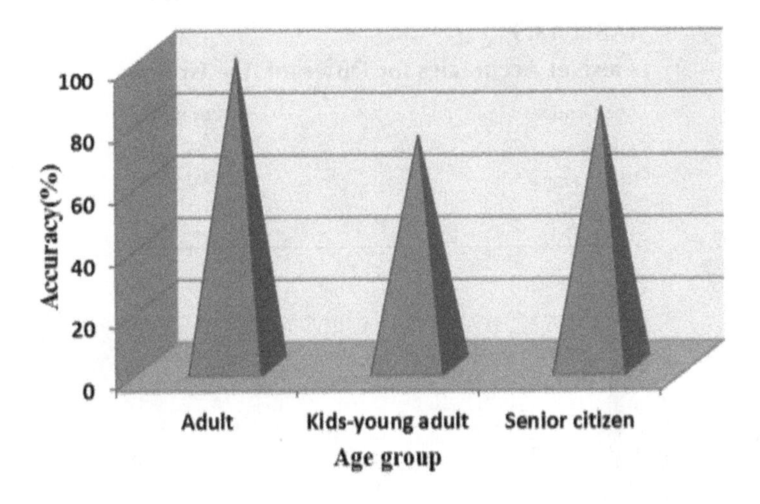

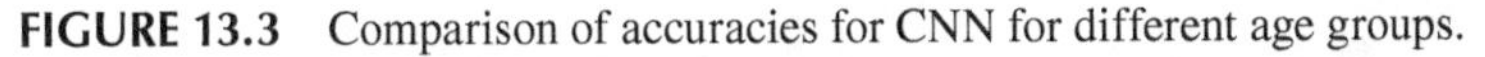

FIGURE 13.3 Comparison of accuracies for CNN for different age groups.

(5) Multiclass SVM was trained on training image features (labeled male/female).
(6) A trained SVM model was used to test age group of images in the test database.

As shown in Table 13.5, all test samples from the senior citizen group were correctly classified, thus providing good accuracy for the senior citizen age group compared to the two other age groups. Figure 13.5 shows the comparison of accuracies for AlexNet for different age groups.

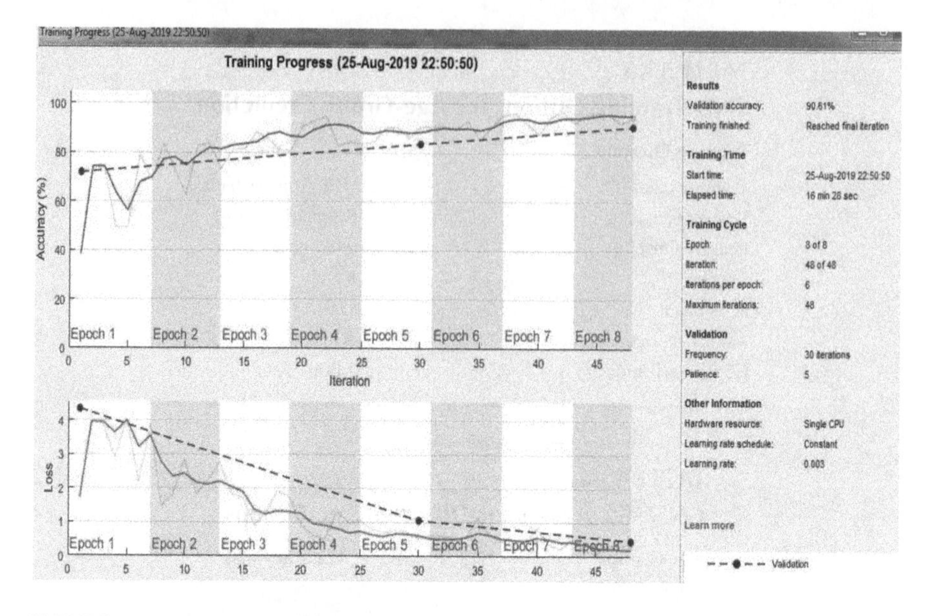

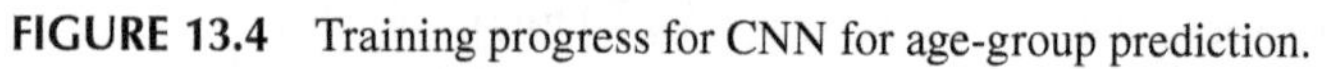

FIGURE 13.4 Training progress for CNN for age-group prediction.

TABLE 13.5
AlexNet Accuracies for Different Age Groups

Age Groups	Accuracy (%)
Adult	96.10
Kids-young adult	90
Senior citizen	100

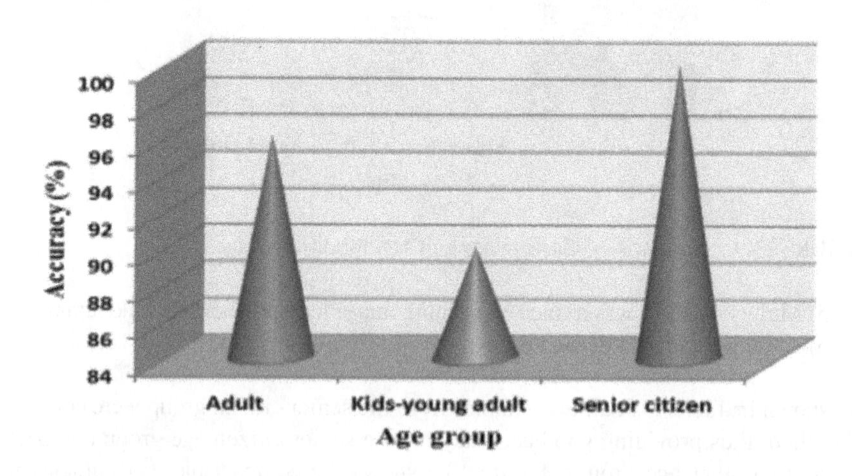

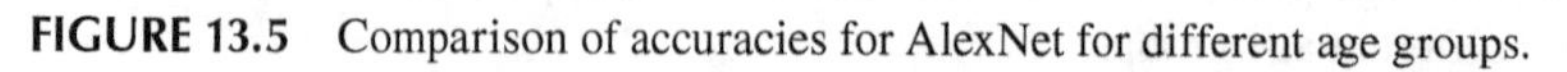

FIGURE 13.5 Comparison of accuracies for AlexNet for different age groups.

C. Feature extraction from GoogLeNet for age-group identification

Steps 1 and 2 from approach 2 were executed in a similar manner as in AlexNet to get 224×224×3 iris image sizes. This was followed by:

- Loading of pre-trained DAG network.
- Features were extracted from the pool5-drop_7×7_s1 layer for training and test images. This layer outputs a four-dimensional feature matrix.
- Four-dimensional feature matrix obtained in step 3 was reshaped to get a two-dimensional matrix of size 855×1,024 for training images features and 214×1,024 for test image features, where 855 and 214 represent the number of training and test images, respectively. Thus, the feature vector size for each input image was 1×1,024.
- A multiclass SVM model was trained on training image features to classify the age group of the input iris image. The trained model was further used to classify the age group of test images.

The GoogLeNet network gave correct predictions for adult class whereas less correct predictions for the other classes. Figure 13.6 shows the comparison of accuracies for GoogLeNet for different age groups.

13.5.3 Performance Comparison of the Proposed System with Earlier Methods for Age-Group Prediction

Table 13.6 describes a comparison of the performance of individual implementation of the proposed algorithms for age-group identification with state-of-the-art methods. Observations show that the state-of-the-art methods used different scanners compared to the proposed method. Our proposed method classifies iris image into one of the three age groups and achieved good accuracy for feature extraction by AlexNet (deep pretrained network). The distinguishing points of our study are: More number of subjects than earlier studies, and subjects with an age range of 3–74 years. (Previous studies included subjects above 18 years only and did not consider kids.)

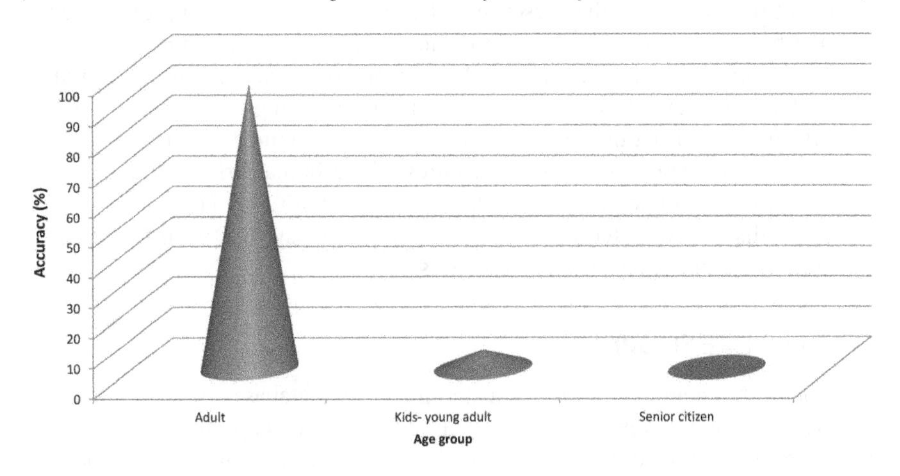

FIGURE 13.6 Comparison of accuracies for GoogLeNet for different age groups.

TABLE 13.6
Comparison of the Proposed System with Earlier Methods for Age-Group Estimation

Methodology	Iris Scanner	Database (Number of Subjects)	Feature Vector Size	Classification	Age Groups	Accuracy (%)
Erbilek 2013	LG Iris Access EOU3000	BMDB (200)	5 geometric features	Sensitivity negotiation method	3	75
Sgroi 2013	LG Iris Access 4000	Own database (100)	630 texture features	Random forest algorithm (Weka 300 trees)	2	64.68
Proposed method (Approach 1)	I Scan 2	Real-time database (213)	No feature extraction	CNN	3	52
Proposed method (Approach 2)	I Scan 2	Real-time database (213)	4,096	AlexNet	3	94.84
Proposed method (Approach 3)	I Scan 2	Real-time database (213)	1,024	GoogLeNet	3	69.01

13.6 CONCLUSIONS

The proposed system is an extension of the existing iris biometrics systems. Existing biometrics reveals only personal identity by acceptance or rejection. The proposed system not only identifies the person but also predicts the age group of the person from the biometric data stored in the database of the system. With the proposed system, an input iris image is classified under one of the three categories: kids to young adult, adult and senior citizen. The system achieved good accuracy (94.84%) with AlexNet-based feature extraction method. The obtained results have further strengthened the hypothesis that the features of the human iris develop from the childhood to the adulthood stage and later diminish with age. This may be due to eye-related diseases or external factors. In the proposed system, we used this information to determine age from iris biometrics.

13.7 FUTURE SCOPE

The proposed system can be implemented for larger databases to achieve the best possible performance for deep neural networks. Future scope for this work may include age estimation for larger datasets with more subjects than those considered in this study. The larger database may help deep networks to reach their best-optimized performance. We are confident that the proposed work will increase the interest among upcoming researchers for advanced studies in the domain of iris biometrics.

REFERENCES

Aslam, T. M., Dhillon, B., and Tan, S. Z. 2009. Iris recognition in the presence of ocular disease. *NCBI*, 6, no.34:489–493.

Browning, K., and Orlans, N. 2014. Biometric aging effects of aging on iris recognition. *MITRE.*

Daugman, J. 1994. Biometric personal identification system based on iris analysis. US patent 5291560A.

Daugman, J. 2001. Statistical richness of visual phase information: Update on recognizing persons by iris patterns. *International Journal of Computer Vision*, 45, no.1:25–38.

Erbilek, M., Fairhurst, M., and Abreu, M. C. D. C. 2013. Age prediction from iris biometrics. *Paper Presented at International Conference on Imaging for Crime Detection and Prevention (ICDP)*, London, UK, 1–5.

Fairhurst, M., and Erbilek, M. 2011. Analysis of physical ageing effects in iris biometrics. *IET Computer Vision, In Special Issue: Future Trends in Biometric Processing*, 5, no.6:358–366.

Fenker, S. P., and Bowyer, K. W. 2012. Analysis of template aging in iris biometrics. *Paper Presented at the IEEE Computer Society Conference on Computer Vision and Pattern Recognition Workshops*, Providence, 45–51.

Glasser, A., and Kasthurirangan, V. 2006. Age related changes in the characteristics of the near pupil response. *Journal of Vision Research*, 46, no.8:1393–1403.

Lanitis, A. 2001. A survey of the effects of aging on biometric identity verification. *International Journal of Biometrics*, 2, no.1:34–52.

Li, Y. H., Huang, P. J., and Juan, Y. 2019. An efficient and robust iris segmentation algorithm using deep learning. In: Marco Anisetti (eds.), *Mobile Information Systems.*

Minaee, S., and Abdolrashidi, A. 2019. Deep iris: iris recognition using a deep learning approach. *Computer Vision and Pattern Recognition.*

Muron, A., and Pospisil, J. 2000. The human iris structure and its usages. *Physica*, 39:87–95.

Rajput, M.R., and Sable, G.S. 2019, June. Deep learning based gender and age estimation from human iris. Available at SSRN: https://ssrn.com/abstract=3576471 or http://dx.doi.org/10.2139/ssrn.3576471

Sgroi, A., Bowyer, K. W., and Flynn, P. J. 2013. The prediction of old and young subjects from iris texture. *Paper Presented at International Conference on Biometrics (ICB)*, Madrid, Spain, 1–5.

Trokielewciz, M., Czajka, A., and Maciejewicz, P. 2020. Post-mortem iris recognition with deep-learning-based image segmentation. *Image and Vision Computing*, 94.

Wildes, R. P. 1997. Iris recognition: An emerging biometric technology. *IEEE*, 85, no.9:1349–1363.

14 Hybrid Intelligent Decision Support Systems to Select the Optimum Fuel Blend in CI Engines

Sakthivel Gnanasekaran and Naveen Kumar P.
Vellore Institute of Technology Chennai

CONTENTS

14.1 INTRODUCTION

Energy plays a vital role in the world's economy. Continuous industrial revolution and rapid population growth have resulted in an energy crisis. Energy contributes significantly in this socioeconomic environment to meet human needs. The time has come to realize that there is a threat to the energy resources needed to meet our daily fuel needs. According to the statistics of energy outlook 2030, globally, the available energy resources will deplete from 2050 to 2075 due to an increase in the rate of energy consumption by 1.6% every year. Therefore, it is necessary to find alternatives to fossil fuels such as petroleum. Biodiesel has emerged as a widely used substitute for fossil fuels as it causes less pollution, preserves the ecosystem and has similar combustion properties of diesel. Recently, many researchers have been working in the field of substitute fuels which can be used without any alterations in the engines. Biodiesel is manufactured by esterification of either vegetable oils or animal fats with alcohol (Vilela et al. 2010). Biodiesel extracted from cultivated crops can pose a threat to food security (Boyd et al. 2004). Meanwhile, the unwanted parts of animals can serve as an excellent source of raw materials for the production of biodiesel. According to the Central Institute of Fisheries Technology (CIFT), every year large quantities of fish waste products are disposed of by various fish food manufacturers. As per the data given by the World Fish Oil and Fishmeal Organization, nearly 0.00101 billion tons of fish oil is produced every year, which is likely to increase up to five times in the next 5 years (Jayasinghe and Hawboldt 2012, Reyes et al. 2012). Hence, biodiesel derived from fish oil needs to be given more attention to be considered as a good alternative energy fuel for diesel, thereby minimizing environmental pollution and reassuring energy security.

Many researchers have investigated fish oil biodiesel in evaluating engine performance characteristics. The results indicate that the engines performed normally with reduced emissions and increased efficiency (Preto et al. 2008, Aryee et al. 2009, Godiganur et al. 2010, Lin and Li 2009a, 2009b, Kato et al. 2004). Carbon monoxide (CO) and carbon dioxide (CO_2) emissions decreased reasonably while using fish oil as biodiesel (Steigers 2002, Behcet 2011, Sakthivel and Nagarajan 2011, Sakthivel et al. 2014). Steigers (2002) and Behcet (2011) studied engine performance using fish oil as fuel by changing fuel combinations at intervals of 20% from 0% to 100%, and the results showed that the engine performed as usual, assuring its suitability as a substitute for petroleum. In this chapter, fish oil is esterified using ethanol and combined with diesel in various combinations to operate an engine at 1,500 rpm. By varying loads and biodiesel–diesel blending ratio, performance characteristics and emission behavior of the diesel engine are studied. Because both blends and loads are varied to examine engine characteristics, identifying the best blending combination to run the Internal combustion (IC) engine is a hard task using a graphical method. It is observed from the literature that several researchers have identified the best blend using conventional graphical methods by primarily considering nitrous oxides, smoke and brake thermal efficiency (BTE) performance (Vedaraman et al. 2011, Shanmugam et al. 2011, Sakthivel 2016). It is observed that all the emission and performance parameters have been accounted for to determine the best blend.

From an environmental viewpoint, the main objective of engine researchers is to minimize fuel consumption by improving engine performance. Multicriteria decision making (MCDM) can be applied to address this complex multiobjective optimization problem. Although the application of MCDM in the automotive industry is gaining importance, it needs more focus for the investigation in the IC engine sector to evaluate suitable biodiesel blends. Poh and Ang (1999) and Rassafi et al. (2006) identified the best alternative using MCDM techniques. Winebrake and Creswick (2003) proposed MCDM to predict the future of hydrogen as a fuel in the transportation system. Perimenis et al. (2011) recommended the decision support system to assess biofuels. Sakthivel et al. (2013) developed an MCDM model to identify the best vehicle among various alternatives. Nwokoagbara et al. (2015) applied decision techniques in biodiesel research. Rathi et al. (2016) suggested an MCDM method in the automotive industry for a six sigma study. Mayyas et al. (2016) and Girubha and Vinodh (2016) identified the best ecomaterial using F-TOPSIS and F-VIKOR approach for vehicle body panels in the automotive industry. However, no study has integrated the MCDM model in IC engines to select the best fuel blends. Hence, TOPSIS, VIKOR, COPRAS-G and MOORA approaches in MCDM can be applied to select the best combination by observing the engine parameters.

The objectives of this chapter are (1) to examine the engine performance, combustion and emission parameters using diesel and biodiesel combinations at various loads and to compare them with diesel; and (2) to consider fuzzy theory integrated with TOPSIS, VIKOR, COPRAS-G and MOORA, multiobjective optimization MCDM techniques, to find the ideal combination at different loads to achieve the best thermal efficiency and low emissions. The performance of each method is compared with other methods and the ranking performance of all methods is validated.

14.2 EXPERIMENTAL SETUP

Experimental investigations was carried out in a single-cylinder compression ignition engine of 4.4kW at constant speed 1500rpm. The braking load was varied by connecting an electrical dynamometer to the engine. A burette and stopwatch were used to compute fuel consumption. The in-cylinder pressure was determined using a piezoelectric transducer mounted on the engine cylinder head. The opening pressure of the injector was set at 200 bar. The temperature of the exhaust gas from the external exhaust was measured by a K-type thermocouple. The heat release rate (HRR) and crank angle pressure were evaluated using a data acquisition system. First, diesel fuel was used to warm the engine for operation, and the successive tests were done at the same speed using various fuel blends. For minimizing errors, a steady-state test was performed three times. AVL437-series smoke meter and AVL444-series emission analyzer were used for measuring the intensity of the exhaust gases. The engine configurations are tabulated in Table 14.1. The overall view of the experimental setup is shown in Figure 14.1. To ensure data accuracy, every test was repeated four times. The observed engine performance is detailed in Table 14.6.

FIGURE 14.1 Photographic view of the experimental setup.

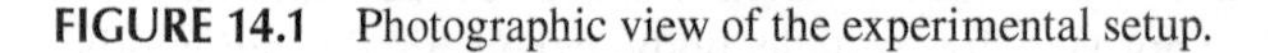

TABLE 14.1
Engine Specifications

Items	Specification
Make	Kirloskar
Cylinder number	1
Type	Four-stroke, stationary, constant speed, direct injection, air cooled, diesel engine
Bore × stroke	80 mm × 110 mm
Displacement	661cc
Compression ratio	17.5:1
Maximum power/speed	4.4 kW/1,500 rpm
Injection timing	24°BTDC
Injection pressure	210 bar

14.3 LITERATURE REVIEW

14.3.1 Fuzzy Logic

Fuzzy logic was described by Zadeh in 1965 to identify the imprecision in real-world decision-making applications. It is an easy-going intelligent system, which mirrors the capability of human decision making to minimize the uncertainty in an ambiguous environment. The principal benefits of the fuzzy logic system are reliability, speed, ease of use and cost. A fuzzy system is created on the system of fuzzy sets, linguistic variables and if–then rules. The basic elements of fuzzy logic are fuzzification, fuzzy inference and defuzzification. The main objective of fuzzification is to convert crisp values to fuzzy linguistic values and levels of the membership function. The state of the membership function may be trapezoidal, triangular, Gaussian, bell- and S-shaped, and the most popular method is the following triangular fuzzification method:

$$\mu(x) = \begin{cases} 0 \text{ if } x \le a \\ \dfrac{x-a}{b-a} \quad \text{if } a < x \le b \\ \dfrac{c-x}{c-b} \quad \text{if } b < x \le c \\ 0 \text{ if } x < c \end{cases} \tag{14.1}$$

Here a and c indicate the minimum and maximum boundaries of fuzzy evaluation.

14.3.2 F-TOPSIS

TOPSIS is one of the methods dealing with the rank reversal technique, and it is fairly easy to identify a suitable alternative using this method. The basic objective of TOPSIS is to compute the distance of the ideal and nonideal solutions by taking into account the distance of other alternatives (Buyukozkan and Guleryuz 2016). Higher performance and lower emissions are considered as the positive ideal solution (PIS), whereas higher emissions and lower performance are considered as the negative ideal solution (NIS) (Etghani et al. 2013). Several investigators have anticipated the use of TOPSIS to identify a suitable alternative in a critical decision-making environment (Cavallaro 2013, Yan et al. 2011, Soufil et al. 2013). Wang recommended TOPSIS to identify the relative preference function (Wang 2014). Chang et al. (2016) used TOPSIS for the evaluation of the diesel engine quality. Chandrasekhar and Raja (2016) applied TOPSIS in automobile engineering to identify suitable materials. The TOPSIS process is as follows:

Step 1: A team of three or more experts (r) is formed to frame the pair-wise comparison matrix, D_r=(e=1, 2,…μ. E), and the decision can be indicated in Traingular Fuzzy Numbers (TFN), that is, B_r=(r=1,2,…, R) with membership function (y). The suitable linguistic variables are also chosen to identify the alternatives by considering the various criteria.

Step 2: The linear scale transformation is used to convert several criteria into an equivalent mode to evaluate the criteria weights.

$$\tilde{R} = \left[\tilde{r}_{pq}\right]_{m \times n}, \; p = 1,2,\ldots,x; q = 1,2,\ldots,y \tag{14.2}$$

Where

$$\tilde{r}_{pq} = \left(\frac{a_{pq}}{\sum_{p=1}^{k} a_k}, \frac{b_{pq}}{\sum_{q=1}^{k} b_k}, \frac{c_{pq}}{\sum_{p=1}^{k} c_k} \right)$$

Step 3: Structure of the fuzzy decision matrix is:

$$\tilde{D} = \begin{bmatrix} \tilde{m}_{11} & \tilde{m}_{12} & \cdots & \tilde{m}_{1y} \\ \tilde{m}_{21} & \tilde{m}_{22} & \cdots & m_{2y} \\ \vdots & \vdots & \cdots & \vdots \\ \tilde{m}_{11} & \tilde{m}_{12} & \cdots & \tilde{m}_{1y} \end{bmatrix}$$

Here, $\tilde{x}_{pq} = (a_{pq}, b_{pq}, c_{pq}); p = 1,2,\ldots,x; q = 1,2,\ldots,y$ is evaluated by positive TFN. The rating of fuzzy and the weight importance of kth selection-maker are $\tilde{x}_{pq} = (a_{pq}, b_{pq}, c_{pq})$ and $\tilde{w}_{pqr} = (w_{qr1}, w_{qr2}, w_{qr3})$

$$a_{pq} = \min_k \{a_{pqr}\}, b_{pq} = \frac{1}{K}\sum\nolimits_{k=1}^{k} b_{pqr}, c_{pq} = \max_k \{c_{pqr}\} \tag{14.3}$$

Step 4: The normalization for the created decision matrix is as follows:

$$g_{pq} = \frac{x_{pq}}{\sqrt{\sum\nolimits_{q=1}^{J} x_{pq}^{2}}} \tag{14.4}$$

Step 5: The weighted normalized decision matrix is the product of the normalized decision matrix r_{ij} and its weight w_p

$$V_{pq} = w * r_{pq} \;\; p = 1,2,\ldots,x; q = 1,2,\ldots,y \tag{14.5}$$

Step 6: Next, the fuzzy PIS (FPIS, H^*) and fuzzy NIS (FNIS, H^-) are identified as:

$$H^* = (\tilde{v}_1^*, \tilde{v}_2^*, \ldots, \tilde{v}_q^*) \tag{14.6}$$

$$H^- = (\tilde{v}_1^-, \tilde{v}_2^-, \ldots, \tilde{v}_q^-) \tag{14.7}$$

$$\tilde{v}_q^* = \max_p \{v_{pq3}\} \text{ and } \tilde{v}_q^- = \min_p \{v_{pq1}\}$$

Step 7: The distance of the PIS and NIS for all alternatives are computed by the Euclidean distance:

$$d_i^* = \sum\nolimits_{q=1}^{n} d_v(\tilde{v}_{pq}, \tilde{v}_q^*), \; p = 1,2,\ldots,x \tag{14.8}$$

$$d_i^- = \sum\nolimits_{q=1}^{n} d_v(\tilde{v}_{pq}, \tilde{v}_q^-), \; p = 1,2,\ldots,x \tag{14.9}$$

Step 8: The relation nearness of the xth alternative with respect to the ideal solution H^+ is defined as:

$$\mathrm{CC}_p = \frac{d_p^-}{d_p^- + d_p^+}, i = 1, 2, \ldots, x \tag{14.10}$$

Step 9: A set of alternatives are ranked prior as per the values in the down direction order of CC_q^*.

14.3.3 F-VIKOR

VIKOR (VlseKriterijumska Optimizacija Kompromisno Resenje), also called compromise ordering rank method, is a conceivable arrangement that is nearest to the perfect arrangement, and the significance of trade-off is created by common concession. VIKOR was introduced by Opricovic in 1998. In view of the idea of fuzzy set hypothesis and VIKOR strategy, the projected-VIK technique is utilized after basic leadership manages a few options which can be positioned regarding any distinct criteria. F-VIKOR is proposed as fuzzy TOPSIS to identify the finest substitute based on the same criteria and linguistic variables. Diakoulaki et al. (1999) used VIKOR for energy analysis. Tzeng et al. (2005) proposed this approach for framing transportation strategy in Singapore. Kaya and Kahraman (2010) applied VIKOR as a result validation tool for the promotion of energy conservation. Cristobal (2011) selected the optimum energy project for Spain using VIKOR. Vucija et al. (2013) analyzed sustainable hydropower using VIKOR technique. Taylan et al. (2016) assessed the energy efficiency processes in the petrochemical industry using VIKOR. Li and Zhao (2016) evaluated the performance of emerging ecoindustrial plants in China using the VIKOR technique to disseminate the sustainable production of coal resources.

Step 1: The framing of pair-wise comparison matrix, fuzzy decision matrix and identifying criteria weights similar to TOPSIS methodology.

Step 2: The normalized matrix is now defuzzified using the center of area method as follows:

$$\mathrm{BNP} = \frac{\left[(z - x) + (y - x)\right]}{3} + x \tag{14.11}$$

Step 3: Next, the normalization of defuzzification matrix is done similar to the TOPSIS methodology

Step 4: After normalization, determine the fuzzy best value and fuzzy worst value as:

$$\tilde{f}_q^+ = \max{}_q \tilde{x}_{pq}, \tilde{f}_q^- = \max{}_p \tilde{x}_{pq} \tag{14.12}$$

Step 5: The utility and regret measure for fuzzy GRA-VIKOR is determined using Equations 14.19 and 14.20

$$S_q = \sum_{p=1}^{n} w_p \left(f_p^* - f_{pq}\right) / \left(f_p^* - f_p^-\right) \quad (14.13)$$

$$R_q = \max_p \left[w_q \left(f_p^* - f_{pq}\right) / \left(f_p^* - f_p^-\right) \right] \quad (14.14)$$

Step 6: The maximum and minimum values of utility and regret measures are identified using Equations 14.21 and 14.22

$$\tilde{S}_p = \sum_{q=1}^{k} \tilde{w}_q \left(\tilde{f}_q^* - \tilde{x}_{pq}\right) / \left(\tilde{f}_q^* - \tilde{f}_q^-\right)$$

$$\tilde{S}^* = \min i_p \tilde{S}_p, \tilde{S}^- = \max_p \tilde{S}_p, \quad (14.15)$$

$$\tilde{R}^* = \min i_p \tilde{R}_p, \tilde{R}^- = \max_p \tilde{R}_p, \quad (14.16)$$

Step 7: Final ranking for each alternative is:

$$\tilde{Q}_p = v\left(\tilde{S}_p - \tilde{S}_p^*\right) / \left(\tilde{S}_p^- - \tilde{S}_p^*\right) + (1-v)\left(\tilde{R}_p - \tilde{R}^*\right) / \left(\tilde{R}^- - \tilde{R}^*\right) \quad (14.17)$$

v is presented as the weight of the maximum group utility. Therefore, the smaller the values of Q_p, the better the alternative.

14.3.4 F-MOORA

MOORA optimization is a multiobjective method introduced by Brauers to deal with difficult decision-making complications (Brauers et al. 2004). Karande and Chakraborty (2012) used the MOORA technique for selecting materials. Uttam Kumar and Sarkar (2012) proposed this method for an intelligent manufacturing system. Archana and Sujatha (2012) suggested this technique to select the best network which reduces energy consumption. Vantansever and Kazancoglu (2014) proposed a manufacturing firm's cutting-machine selection problem using Fuzzy AHP and MOORA. Dincer (2015) applied this method for a stock selection approach in banking. Uygurturk (2015) evaluated internet branches of banks using this technique. Kundakci (2016) proposed MOORA for an automobile selection problem in Turkey.

Following are the steps of fuzzy MOORA procedure :

Step 1: The framing of pair-wise comparison matrix, fuzzy decision matrix and identifying criteria weights are similar to TOPSIS methodology.

Step 2: For multiobjective optimization, the parameter y_p decides the best alternative as:

$$y_p = \sum_{q=1}^{g} \tilde{v}_{pq} - \sum_{q=g+1}^{g} \tilde{v}_{pq} \quad (14.18)$$

Where maximum number of criteria is taken as g, minimum number of criteria is considered as $(n - g)$ and y_p is the normalized calculation value of pth alternate with respect to all criteria. The y_p value can be positive or negative depending upon the maximizing or minimizing criteria. The higher the value of y_p, the better is the alternative.

14.3.5 COPRAS-G

Deng (1982) designed the Grey system theory. According to Deng, the merits of Grey relational study are as follows:

- It contains basic calculations and needs fewer samples
- A peculiar allotment is not required
- The results following the Grey relational ranking do not contradict with qualitative analysis.

The Grey model is efficient in dealing with distinct data statistics as it is a transfer function model. The COPRAS-G technique involves actual preconditions of choice generating and the use of Grey logics theory. Zavadskas et al. (2008) proposed the COPRAS-G technique for selecting project managers. Chatterjee and Chakraborty applied COPRAS-G to decide the problem of material selecting for cost-effective manufacturing. Madjid et al. (2013) used this technique to select a social media platform. Liou et al. (2015) proposed COPRAS-G for filtering and picking providers in green supply chain management.

The COPRAS-G technique is as follows:

Step 1: The framing of pair-wise comparison matrix, uncertain decision matrix and identifying criteria weights are similar to the TOPSIS methodology.

Step 2: The relative significance of each alternative is determined by P_q value, which indicates the sum of criterion values (greater is better) defined as follows:

$$\frac{1}{2}\sum_{p=1}^{d}\left(\tilde{a}_{pq}+\tilde{c}_{pq}\right) \tag{14.19}$$

Where d is the number of attributes that need to be increased.

R_q value indicates the sum of criterion values (smaller is better) defined as follows:

$$\frac{1}{2}\sum_{p=d+1}^{e}\left(\tilde{a}_{pq}+\tilde{c}_{pq}\right) \tag{14.20}$$

Where d and e are the number of alternatives that need to be minimized.

Minimum value of R_q is defined as follows:

$$R_{\min} = \mathrm{Min}_q\left(R_q\right) \tag{14.21}$$

The relative weight of each alternatives (Q_q) is determined by:

$$Q_q = P_q + \frac{\sum_{q=1}^{n} R_q}{R_q \sum_{q=1}^{n} \frac{1}{R_q}} \tag{14.22}$$

Step 3: To determine the utility degree of each alternative, the optimality criterion K should be defined as follows:

$$K = \text{Max}_q \left(Q_q\right) \tag{14.23}$$

The utility degree is determined by comparing with others. The utility value varies from 0% (worst alternative) to 100% (good alternative). The utility degree is calculated as follows:

$$N_q = \frac{Q_q}{K} * 100 \tag{14.24}$$

Where Q_q and K are the significances of alternatives.

14.4 THE PROPOSED METHOD

The proposed approach comprises the following stages and is described in Figure 14.2.

(1) Engine operation and analysis of results
(2) Fuzzy computations
(3) To identify the optimum blend using TOPSIS, VIKOR, MOORA and COPRAS-G.

Exploratory analysis was done using a variable load CI engine with rated power at a defined speed to observe the emission behavior and engine performance. Next, the fuzzy logic was trained to frame the pair-wise comparison matrix to identify the

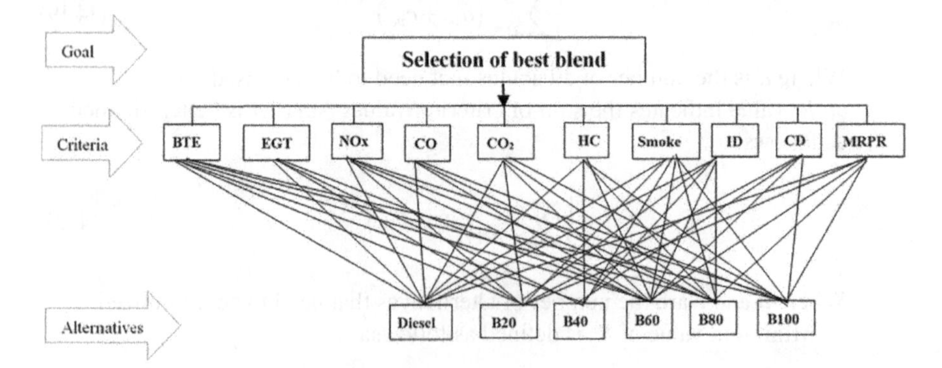

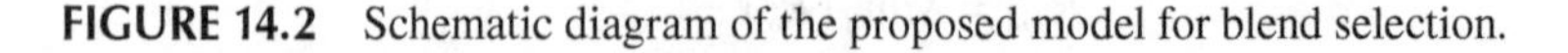
FIGURE 14.2 Schematic diagram of the proposed model for blend selection.

criteria weights. F-TOPSIS, F-VIKOR, F-MOORA and COPRAS-G were used for ranking the substitutes using experiential evaluations and calculated criteria weights.

14.4.1 Selection Criteria for Best Combination

In this investigation, the selection criteria for the best blend were finalized from the combined opinion of researchers (Lin et al. 2009a, Stiegers 2002, Shanmugam et al. 2011)[]. They are defined as follows:

(1) Nitrogen oxide (NOx): The oxygen concentration, in-cylinder temperature, ignition delay and residence time of the combustion flame are the key aspects that contribute to the formation of NOx.
(2) Smoke: Smoke development mainly depends on the availability of oxygen, mode of the mixture and the thermal breaking of unburned hydrocarbon in the rich fuel section. The exhaust smoke is a known indicator of combustion quality of the engine.
(3) BTE: It denotes the efficiency with which the thermodynamic input is transformed into mechanical work. It is defined as the output brake power as a function of chemical energy in the petroleum source.
(4) CO_2: CO_2 assists as a heat-absorbing agent due to higher heat capacity and indicates how efficiently the fuel is burned in the combustion chamber.
(5) CO: CO is formed as a result of incomplete combustion due to insufficient oxygen in the air–fuel mixture ratio.
(6) Hydrocarbon: Hydrocarbons along with oxygen in the fuel are combusted and the unburned hydrocarbons are released as hydrocarbon emission.
(7) Exhaust gas temperature: It specifies the temperature of the tested fuel during the flaming period.
(8) Ignition delay: It regulates the amount of premixed fire, and depends on the amount of pressure rise, peak pressure, engine noise and vibrations.
(9) Combustion duration: It is defined as the duration from the start to the end of the combustion cycle and it can be determined from HRR
(10) Maximum rate of pressure rise: It differs according to the quantity of fuel contributing in the premixed combustion phase during the early stage.

14.5 RESULTS AND DISCUSSION

A decision hierarchy was framed using selected parameters and the fuel alternatives, as shown in Figure 14.3. The comparative weightings were computed using the pair-wise matrix according to Satty's scale. The matrix was formed as per the opinion of experts and industrialists using a questionnaire survey.

14.5.1 F-TOPSIS Computations

The engine characteristics at 75% loads are discussed to demonstrate the F-TOPSIS computations. The pair-wise matrix of engine parameters was derived using equation 14.2 and is detailed in Table 14.2.

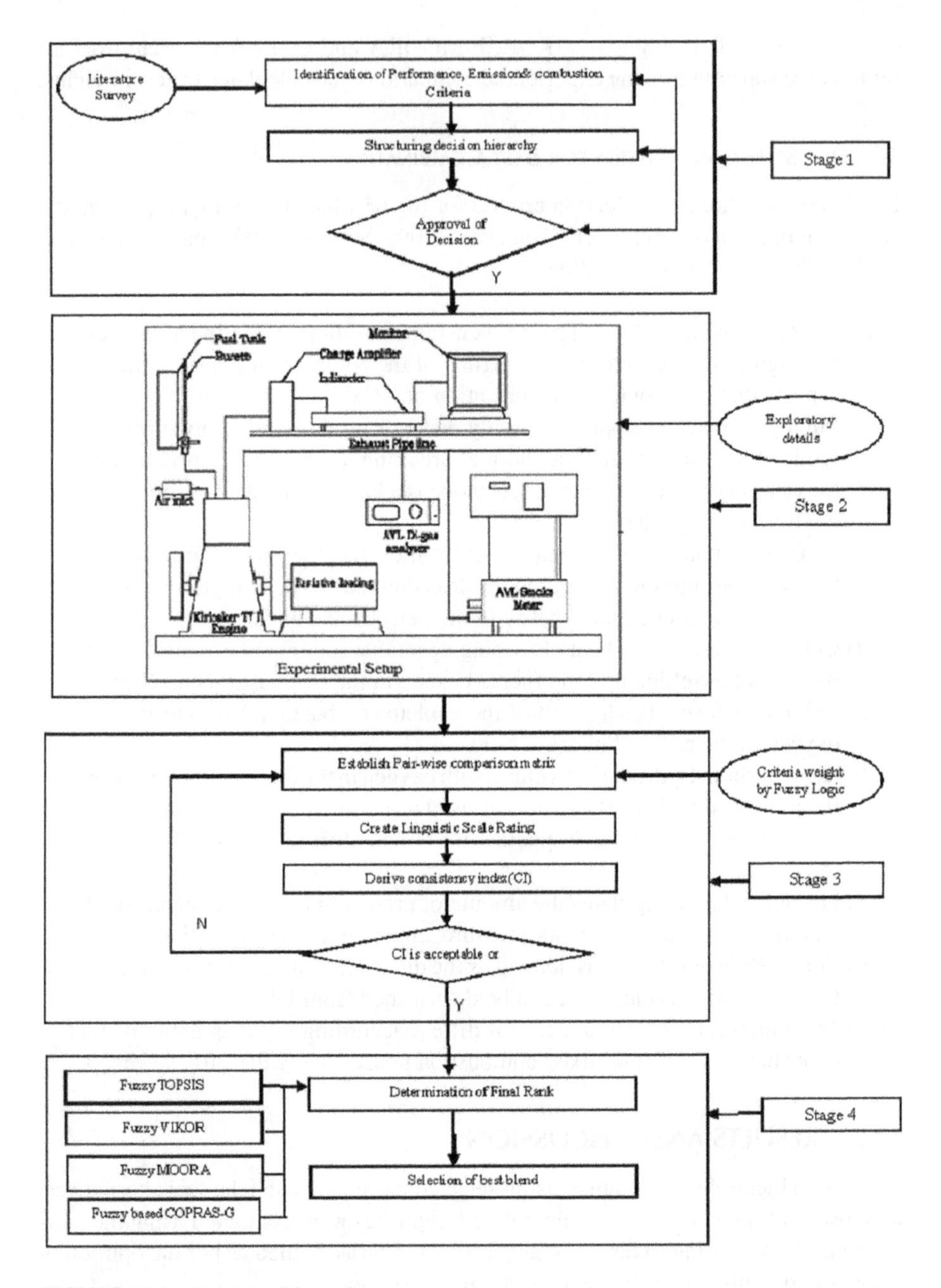

FIGURE 14.3 Decision hierarchy.

The conclusion matrix and weighted normalized matrix are framed by Equations 14.3–14.5 using fuzzy weight, which is tabulated in Table 14.3.

Their corresponding weights are also shown in Tables 14.4 and 14.5.

An investigative study of engine performance parameters for dissimilar blends and loads are detailed in Table 14.6.

TABLE 14.2
Decision-Maker Matrix for Criteria

Criteria	NO_x	Smoke	BTE	CO_2	CO	HC	EGT	ID	CD	MRPR
NO_x	–	MH	MH	H	H	H	VH	VH	VH	VH
Smoke	ML	–	M	MH	H	H	VH	VH	VH	VH
BTE	ML	M	–	MH	MH	H	VH	VH	VH	VH
CO_2	L	ML	ML	–	M	MH	H	H	VH	VH
CO	L	L	ML	M	–	M	MH	MH	H	VH
HC	L	L	L	ML	M	–	M	M	MH	H
EGT	VL	VL	VL	L	ML	M	–	M	M	MH
ID	VL	VL	VL	L	ML	M	M	–	M	M
CD	VL	VL	VL	VL	L	ML	M	M	–	M
MRPR	VL	VL	VL	VL	VL	L	ML	M	M	–

TABLE 14.3
Linguistic Fuzzy Criterion Rating

Rating	*a*	*b*	*c*
VL	0.1	0.1	0.25
ML	0.125	0.25	0.375
L	0.25	0.375	0.5
F	0.375	0.5	0.625
H	0.5	0.625	0.75
MH	0.625	0.75	0.875
VH	0.75	1	1

TABLE 14.4
Weights for Criteria

Criteria	*a*	*b*	*c*
NO_X	5.875	7.5	8.125
Smoke	5.375	7	7.625
BTE	5.25	6.875	7.5
CO_2	4.25	5.625	6.5
CO	3.625	4.875	5.875
HC	2.875	4	5.125
EGT	2.3	3.05	4.25
ID	2.175	2.925	4.125
CD	1.9	2.525	3.75
MRPR	1.625	2.125	3.375

TABLE 14.5
Normalized Weights for Criteria

Criterion	Weights *a*	*b*	*c*
NO_X	0.10444	0.16129	0.2305
Smoke	0.09556	0.15054	0.21631
BTE	0.09333	0.14785	0.21277
CO_2	0.07556	0.12097	0.1844
CO_2	0.06444	0.10484	0.16667
HC	0.05111	0.08602	0.14539
EGT	0.04089	0.06559	0.12057
ID	0.03867	0.0629	0.11702
CD	0.03378	0.0543	0.10638
MRPR	0.02889	0.0457	0.09574

The change of observed performance into seven stages of fuzzy linguistics is tabularized in Table 14.7.

The conclusion matrix, fuzzy decision matrix and weighted matrix was framed using a fuzzy linguistics concept, as shown in Tables 14.8–14.11.

Next, the ideal and nonideal solutions for the alternatives were determined using Equations 14.6 and 14.7 for various loads and detailed in Table 14.12. The computation of NO_x at 75% load was as follows:

$$H^* = \left\{\begin{matrix} (0.04268, 0.04268, 0.04251, 0.04234, 0.04216, 0.04216), \\ (0.06634, 0.06634, 0.06598, 0.06571, 0.06544, 0.06526), \\ (0.09506, 0.09506, 0.09487, 0.09448, 0.0941, 0.09391) \end{matrix}\right\}$$

$$= \{0.04216, 0.06526, 0.09391\}$$

$$H^- = \left\{\begin{matrix} (0.04268, 0.04268, 0.04251, 0.04234, 0.04216, 0.04216), \\ (0.06634, 0.06634, 0.06598, 0.06571, 0.06544, 0.06526), \\ (0.09506, 0.09506, 0.09487, 0.09448, 0.0941, 0.09391) \end{matrix}\right\}$$

$$= \{0.04268, 0.06634, 0.09506\}$$

The PIS and NIS were calculated for the remaining loads using the same process. The distance of the alternative from the ideal and nonideal solution was calculated using Equations 14.8 and 14.9 and tabularized in Table 14.13. The computation for B60 alternative D–(B60, IS) and D^*(B60, NIS) for 75% load are:

TABLE 14.6
Experimental Performance and Emission Readings Observed from Engine for Various Alternative Blends

Criteria / Load	Blends	NOx (ppm)	Smoke (%)	BTE (%)	CO_2 (%vol)	CO (%vol)	HC (ppm)	EGT (C)	ID (CA)	CD (CA)	MRPR (bar/CA)
0%	Diesel	239	5	0	2.2	0.07	29	157	16	46.28	4.55
	B20	221	10.1	0	2	0.06	27	164	15.9	45.16	4.01
	B40	219	12.4	0	2	0.06	26	165	15.82	43.08	4.01
	B60	201	14.2	0	2	0.05	25	168	15.44	44.44	3.48
	B80	191	11.8	0	2	0.05	24	166	15.42	42.68	3.48
	B100	178	20.5	0	2	0.04	23	165	15.34	41.135	3.09
25%	Diesel	5 ×19	15.2	17.31	3.2	0.08	35	204	15.16	49.3	5.95
	B20	522	16	18.03	3.2	0.07	31	205	14.64	47.26	5.61
	B40	520	16.3	16.32	3.3	0.07	27	211	14.48	44.78	4.81
	B60	518	19.4	16.3	3.2	0.06	26	212	14.48	45.95	4.08
	B80	508	20.3	15.69	3.3	0.06	26	204	14.28	44.74	4.01
	B100	492	25.7	14.54	3.5	0.05	24	215	14.02	41.44	3.38
50%	Diesel	986	18.5	26.17	4.8	0.09	38	266	14.22	52.14	6.42
	B20	987	18	27.05	4.5	0.08	34	266	13.64	50.52	5.61
	B40	964	18.6	24.49	4.5	0.08	29	270	13.52	47.38	4.81
	B60	944	23.6	23.99	4.7	0.07	31	270	13.42	48.46	4.4
	B80	935	23.4	21.73	4.7	0.07	29	262	13.24	45.24	4.28
	B100	904	31.4	22.55	4.9	0.06	28	272	12.84	47.12	3.8

(*Continued*)

TABLE 14.6 (*Continued*)
Experimental Performance and Emission Readings Observed from Engine for Various Alternative Blends

Criteria / Load	Blends	NOx (ppm)	Smoke (%)	BTE (%)	CO_2 (%vol)	CO (%vol)	HC (ppm)	EGT (C)	ID (CA)	CD (CA)	MRPR (bar/CA)
75%	Diesel	1,357	25.8	30.71	6.2	0.09	37	321	13.72	53.28	6.42
	B20	1,358	22.5	30.91	5.8	0.09	35	325	13.52	52.66	5.68
	B40	1,351	23.9	30.41	6	0.08	33	330	13.46	48.28	5.35
	B60	1,346	29.8	27.68	6.2	0.08	31	327	13.42	49.73	5.08
	B80	1,340	28.8	27.16	6.2	0.07	29	332	12.66	51.76	4.55
	B100	1,336	37.4	26.71	6.4	0.07	31	340	11.14	50.34	3.7
100%	Diesel	1,700	32	32.63	8.2	0.11	40	398	13.42	54.81	6.68
	B20	1,689	33.3	34.78	7.4	0.09	37	392	13.38	53.42	6.15
	B40	1,666	36	33.91	7.7	0.09	35	403	13.3	52.22	5.61
	B60	1,642	41.3	31.99	7.7	0.08	32	398	13.04	50.55	5.35
	B80	1,642	41.6	30.17	7.7	0.08	33	397	11.68	49.34	4.55
	B100	1,606	50.9	29.68	8.2	0.07	32	399	10.92	48.94	3.9

TABLE 14.7
Linguistic Ratings for Alternatives for 75% Load

	NO_X			Smoke			BTE		
Rating	*a*	*b*	*c*	*a*	*b*	*c*	*a*	*b*	*c*
VP	1336	1336	1341.5	22.5	22.5	26.225	26.71	26.71	27.76
P	1338.75	1341.5	1344.25	24.3625	26.225	28.0875	27.235	27.76	28.285
MP	1341.5	1344.25	1347	26.225	28.0875	29.95	27.76	28.285	28.81
F	1344.25	1347	1349.75	28.0875	29.95	31.8125	28.285	28.81	29.335
MG	1347	1349.75	1352.5	29.95	31.8125	33.675	28.81	29.335	29.86
G	1349.75	1352.5	1355.25	31.8125	33.675	35.5375	29.335	29.86	30.385
VG	1352.5	1358	1358	33.675	37.4	37.4	29.86	30.91	30.91

	CO_2			CO			HC		
Rating	*a*	*b*	*c*	*a*	*b*	*c*	*a*	*b*	*c*
VP	5.8	5.8	5.95	0.07	0.07	0.075	29	29	31
P	5.875	5.95	6.025	0.0725	0.075	0.0775	30	31	32
MP	5.95	6.025	6.1	0.075	0.0775	0.08	31	32	33
F	6.025	6.1	6.175	0.0775	0.08	0.0825	32	33	34
MG	6.1	6.175	6.25	0.08	0.0825	0.085	33	34	35
G	6.175	6.25	6.325	0.0825	0.085	0.0875	34	35	36
VG	6.25	6.4	6.4	0.085	0.09	0.09	35	37	37

	ID			CD			MRPR		
Rating	*a*	*b*	*c*	*a*	*b*	*c*	*a*	*b*	*c*
VP	321	321	325.75	11.14	11.14	11.785	48.28	48.28	49.53
P	323.375	325.75	328.125	11.4625	11.785	12.1075	48.905	49.53	50.155
MP	325.75	328.125	330.5	11.785	12.1075	12.43	49.53	50.155	50.78
F	328.125	330.5	332.875	12.1075	12.43	12.7525	50.155	50.78	51.405
MG	330.5	332.875	335.25	12.43	12.7525	13.075	50.78	51.405	52.03
G	332.875	335.25	337.625	12.7525	13.075	13.3975	51.405	52.03	52.655
VG	335.25	340	340	13.075	13.72	13.72	52.03	53.28	53.28

	MRPR		
Rating	*a*	*b*	*c*
VP	3.7	3.7	4.38
P	4.04	4.38	4.72
MP	4.38	4.72	5.06
F	4.72	5.06	5.4
MG	5.06	5.4	5.74
G	5.4	5.74	6.08
VG	5.74	6.42	6.42

TABLE 14.8
Augmented Ratings for Alternatives for 75% Load

Criteria	Alternatives				Criteria	Alternatives			
NO_X		**D1**	**D2**	**D3**	**HC**		**D1**	**D2**	**D3**
	Diesel	VG	VG	VG		Diesel	VG	VG	VG
	B20	VG	VG	VG		B 20	MG	G	VG
	B40	MG	MG	G		B 40	MP	MG	F
	B60	MP	MP	F		B60	VP	P	MP
	B80	P	P	VP		B80	VP	VP	VP
	B100	VP	VP	VP		B100	VP	P	MP
Smoke		**D1**	**D2**	**D3**	**EGT**		**D1**	**D2**	**D3**
	Diesel	VP	P	VP		Diesel	VP	VP	VP
	B20	VP	VP	VP		B 20	P	P	VP
	B40	VP	VP	VP		B 40	MP	F	MP
	B60	MP	MP	F		B60	P	P	MP
	B80	MP	MP	F		B80	F	F	MG
	B100	VG	VG	VG		B100	VG	VG	VG
BTE		**D1**	**D2**	**D3**	**ID**		**D1**	**D2**	**D3**
	Diesel	VG	VG	VG		Diesel	VG	VG	VG
	B20	VG	VG	VG		B 20	VG	VG	VG
	B40	VG	VG	VG		B 40	VG	VG	VG
	B60	VP	VP	P		B60	VG	VG	VG
	B80	VP	VP	VP		B80	F	F	MG
	B100	VP	VP	VP		B100	VP	VP	VP
CO_2		**D1**	**D2**	**D3**	**CD**		**D1**	**D2**	**D3**
	DIESEL	MG	MG	G		DIESEL	VG	VG	VG
	B20	VP	VP	VP		B20	VG	VG	VG
	B40	MP	MP	P		B40	VP	VP	VP
	B60	MG	MG	G		B60	P	MP	P
	B80	MG	MG	G		B80	MG	G	MG
	B100	VG	VG	VG		B100	MP	F	MP
CO		**D1**	**D2**	**D3**	**MRPR**		**D1**	**D2**	**D3**
	Diesel	VG	VG	VG		Diesel	VG	VG	VG
	B20	VG	VG	VG		B20	MG	G	MG
	B40	MP	MG	F		B40	F	F	MG
	B60	MP	MG	F		B60	F	F	MG
	B80	VP	VP	VP		B80	P	P	MP
	B100	VP	VP	VP		B100	VP	VP	VP

TABLE 14.9
Fuzzy Decision Matrix for 75% Load

Criteria	NO_X			Smoke			BTE		
Blends	*A*	*b*	*c*	*a*	*b*	*c*	*a*	*b*	*c*
Diesel	1352.5	1358	1358	22.5	23.74167	28.0875	29.86	30.91	30.91
B20	1352.5	1358	1358	22.5	22.5	26.225	29.86	30.91	30.91
B40	1347	1350.667	1355.25	22.5	22.5	26.225	29.86	30.91	30.91
B60	1341.5	1345.167	1349.75	26.225	28.70833	31.8125	26.71	27.06	28.285
B80	1336	1339.667	1344.25	26.225	28.70833	31.8125	26.71	26.71	27.76
B100	1336	1336	1341.5	33.675	37.4	37.4	26.71	26.71	27.76

Criteria	CO_2			CO			HC		
Blends	*a*	*b*	*c*	*a*	*b*	*c*	*a*	*b*	*c*
Diesel	6.1	6.2	6.325	0.085	0.09	0.09	35	37	37
B20	5.8	5.8	5.95	0.085	0.09	0.09	33	35.33333	37
B40	5.875	6	6.1	0.075	0.08	0.085	31	33	35
B60	6.1	6.2	6.325	0.075	0.08	0.085	29	30.66667	33
B80	6.1	6.2	6.325	0.07	0.07	0.075	29	29	31
B100	6.25	6.4	6.4	0.07	0.07	0.075	29	30.66667	33

Criteria	EGT			ID			CD		
Blends	*a*	*b*	*c*	*a*	*b*	*c*	*a*	*b*	*c*
Diesel	321	321	325.75	13.075	13.72	13.72	52.03	53.28	53.28
B20	321	324.1667	328.125	13.075	13.72	13.72	52.03	53.28	53.28
B40	325.75	328.9167	332.875	13.075	13.72	13.72	48.28	48.28	49.53
B60	323.375	326.5417	330.5	13.075	13.72	13.72	48.905	49.73833	50.78
B80	328.125	331.2917	335.25	12.1075	12.5375	13.075	50.78	51.61333	52.655
B100	335.25	340	340	11.14	11.14	11.785	49.53	50.36333	51.405

Criterion	MRPR		
Blends	*a*	*b*	*c*
Diesel	5.74	6.42	6.42
B20	5.06	5.51333	6.08
B40	4.72	5.17333	5.74
B60	4.72	5.17333	5.74
B80	4.04	4.49333	5.06
B100	3.7	3.7	4.38

TABLE 14.10
Normalized Fuzzy Decision Matrix for 75% Load

Criteria	NO_X			Smoke			BTE		
Blends	*a*	*b*	*c*	*a*	*b*	*c*	*a*	*b*	*c*
Diesel	0.408659	0.411294	0.412419	0.301012	0.34917	0.442542	0.413783	0.436028	0.445446
B20	0.408659	0.411294	0.412419	0.301012	0.330909	0.413197	0.413783	0.436028	0.445446
B40	0.406997	0.409073	0.411584	0.301012	0.330909	0.413197	0.413783	0.436028	0.445446
B60	0.405335	0.407407	0.409913	0.350846	0.422215	0.501232	0.370132	0.381719	0.407617
B80	0.403674	0.405742	0.408243	0.350846	0.422215	0.501232	0.370132	0.376782	0.400051
B100	0.403674	0.404631	0.407408	0.450515	0.550044	0.589268	0.370132	0.376782	0.400051

Criteria	CO_2			CO			HC		
Blends	*a*	*b*	*c*	*a*	*b*	*c*	*a*	*b*	*c*
Diesel	0.39912	0.412491	0.427552	0.415252	0.456906	0.47767	0.415315	0.46148	0.485918
B20	0.379491	0.385879	0.402203	0.415252	0.456906	0.47767	0.391583	0.440692	0.485918
B40	0.384398	0.399185	0.412343	0.366399	0.406138	0.451133	0.367851	0.41159	0.459652
B60	0.39912	0.412491	0.427552	0.366399	0.406138	0.451133	0.344118	0.382488	0.433386
B80	0.39912	0.412491	0.427552	0.341972	0.355371	0.398059	0.344118	0.3617	0.40712
B100	0.408934	0.425797	0.432622	0.341972	0.355371	0.398059	0.344118	0.382488	0.433386

Criteria	EGT			ID			CD		
Blends	*a*	*b*	*c*	*a*	*b*	*c*	*a*	*b*	*c*
Diesel	0.394584	0.398675	0.408201	0.401067	0.426622	0.444082	0.409744	0.425454	0.432603
B20	0.394584	0.402608	0.411177	0.401067	0.426622	0.444082	0.409744	0.425454	0.432603
B40	0.400423	0.408507	0.41713	0.401067	0.426622	0.444082	0.380212	0.385528	0.402155
B60	0.397503	0.405557	0.414154	0.401067	0.426622	0.444082	0.385134	0.397173	0.412305
B80	0.403342	0.411457	0.420106	0.37139	0.389852	0.423205	0.3999	0.412145	0.427528
B100	0.4121	0.422272	0.426058	0.341712	0.346397	0.381451	0.390056	0.402164	0.417379

Criteria	MRPR		
Blends	*a*	*b*	*c*
Diesel	0.417672	0.509107	0.556434
B20	0.368191	0.437208	0.526966
B40	0.343451	0.410246	0.497497
B60	0.343451	0.410246	0.497497
B80	0.293971	0.356322	0.438561
B100	0.269231	0.293411	0.379624

TABLE 14.11
Weighted Normalized Fuzzy Decision Matrix for 75% Load

Criteria	NO_X			Smoke			BTE		
Blends	*a*	*b*	*c*	*a*	*b*	*c*	*a*	*b*	*c*
Diesel	0.042682	0.066338	0.095061	0.028763	0.052563	0.095727	0.03862	0.064467	0.094776
B20	0.042682	0.066338	0.095061	0.028763	0.049814	0.089379	0.03862	0.064467	0.094776
B40	0.042509	0.06598	0.094869	0.028763	0.049814	0.089379	0.03862	0.064467	0.094776
B60	0.042335	0.065711	0.094484	0.033525	0.063559	0.108423	0.034546	0.056437	0.086727
B80	0.042161	0.065442	0.094099	0.033525	0.063559	0.108423	0.034546	0.055707	0.085117
B100	0.042161	0.065263	0.093906	0.043049	0.082802	0.127466	0.034546	0.055707	0.085117

Criteria	CO_2			CO			HC		
Blends	*a*	*b*	*c*	*a*	*b*	*c*	*a*	*b*	*c*
Diesel	0.030156	0.049898	0.078839	0.026761	0.047901	0.079612	0.021227	0.039697	0.070648
B20	0.028673	0.046679	0.074165	0.026761	0.047901	0.079612	0.020014	0.037909	0.070648
B40	0.029043	0.048288	0.076035	0.023612	0.042579	0.075189	0.018801	0.035406	0.066829
B60	0.030156	0.049898	0.078839	0.023612	0.042579	0.075189	0.017588	0.032902	0.06301
B80	0.030156	0.049898	0.078839	0.022038	0.037257	0.066343	0.017588	0.031114	0.059191
B100	0.030897	0.051508	0.079774	0.022038	0.037257	0.066343	0.017588	0.032902	0.06301

Criteria	EGT			ID			CD		
Blends	*a*	*b*	*c*	*a*	*b*	*c*	*a*	*b*	*c*
Diesel	0.016134	0.02615	0.049216	0.015508	0.026836	0.051967	0.01384	0.023103	0.046022
B20	0.016134	0.026408	0.049575	0.015508	0.026836	0.051967	0.01384	0.023103	0.046022
B40	0.016373	0.026795	0.050292	0.015508	0.026836	0.051967	0.012843	0.020935	0.042782
B60	0.016253	0.026601	0.049933	0.015508	0.026836	0.051967	0.013009	0.021567	0.043862
B80	0.016492	0.026988	0.050651	0.01436	0.024523	0.049524	0.013508	0.02238	0.045482
B100	0.01685	0.027697	0.051369	0.013213	0.021789	0.044638	0.013175	0.021838	0.044402

Criteria	MRPR		
Blends	*a*	*b*	*c*
Diesel	0.012066	0.023266	0.053276
B20	0.010637	0.01998	0.050454
B40	0.009922	0.018748	0.047633
B60	0.009922	0.018748	0.047633
B80	0.008492	0.016284	0.04199
B100	0.007778	0.013409	0.036347

TABLE 14.12
FPIS and FNIS Computations for 75% Load

	A^-			A^*		
Criteria	*a*	*b*	*c*	*a*	*b*	*c*
NO_X	0.042682	0.066338	0.095061	0.042161	0.065263	0.093906
Smoke	0.043049	0.082802	0.127466	0.028763	0.049814	0.089379
BTE	0.034546	0.055707	0.085117	0.03862	0.064467	0.094776
CO_2	0.030897	0.051508	0.079774	0.028673	0.046679	0.074165
CO	0.026761	0.047901	0.079612	0.022038	0.037257	0.066343
HC	0.021227	0.039697	0.070648	0.017588	0.031114	0.059191
EGT	0.01685	0.027697	0.051369	0.016134	0.02615	0.049216
ID	0.015508	0.026836	0.051967	0.013213	0.021789	0.044638
CD	0.01384	0.023103	0.046022	0.012843	0.020935	0.042782
MRPR	0.007778	0.013409	0.036347	0.012066	0.023266	0.053276

$d^*\left(\text{FPIS}, \text{B60}\right)$

$$= \sqrt{\begin{matrix} \left[\frac{1}{3}\left[(0.04234-0.04268)^2+(0.06571-0.06634)^2+(0.09448-0.09506)^2\right]\right]+ \\ \left[\frac{1}{3}\left[(0.03353-0.04305)^2+(0.06356-0.0828)^2+(0.10842-0.12747)^2\right]\right]+ \\ \left[\frac{1}{3}\left[(0.03455-0.03455)^2+(0.05644-0.05571)^2+(0.08673-0.08512)^2\right]\right]+ \\ \left[\frac{1}{3}\left[(0.03016-0.0309)^2+(0.0499-0.05151)^2+(0.07884-0.07977)^2\right]\right]+ \\ \left[\frac{1}{3}\left[(0.02361-0.002676)^2+(0.04258-0.0479)^2+(0.07519-0.07961)^2\right]\right]+ \\ \left[\frac{1}{3}\left[(0.01759-0.02123)^2+(0.0329-0.0397)^2+(0.06301-0.07065)^2\right]\right]+ \\ \left[\frac{1}{3}\left[(0.01625-0.01685)^2+(0.0266-0.0277)^2+(0.04993-0.05137)^2\right]\right]+ \\ \left[\frac{1}{3}\left[(0.01551-0.01551)^2+(0.02684-0.02684)^2+(0.05197-0.05197)^2\right]\right]+ \\ \left[\frac{1}{3}\left[(0.01301-0.01384)^2+(0.02157-0.0231)^2+(0.04386-0.04602)^2\right]\right]+ \\ \left[\frac{1}{3}\left[(0.00992-0.00778)^2+(0.01875-0.01341)^2+(0.04763-0.03635)^2\right]\right]+ \end{matrix}}$$

$= 0.02399$

TABLE 14.13
D^+ and D^- Computations

Load	0%		25%		50%		75%		100%	
Blends	D^+	D^-	D^+	D^-	D^+	D^-	D^+	D^-	D^+	D^-
Diesel	0.056603	0.086567	0.030243	0.162634	0.044373	0.057171	0.034689	0.048827	0.045702	0.044319
B20	0.069344	0.07515	0.023433	0.170131	0.03278	0.068835	0.023693	0.059906	0.029906	0.060887
B40	0.077782	0.066687	0.026511	0.166491	0.030988	0.070481	0.029485	0.05431	0.034954	0.05599
B60	0.064992	0.07871	0.038892	0.154444	0.048668	0.052789	0.043974	0.039948	0.035073	0.055962
B80	0.055251	0.088745	0.043163	0.150082	0.041672	0.059535	0.039286	0.044168	0.041375	0.049359
B100	0.08615	0.057053	0.1673	0.027435	0.061505	0.039506	0.059272	0.023969	0.054518	0.03543

$$d^{*}\,(\text{FPIS}, \text{B60}) = 0.05990$$

The closeness coefficients were derived from ideal and nonideal solutions using Equation 14.10. The below calculations are for 75% load for the diesel. Finally, the alternatives are ranked as per the closeness value.

$$\text{CC}_1 = \frac{0.05990}{0.0599 + 0.02399} = 0.716$$

A similar computation was done for 25%, 50%, 100% and no-load conditions, and the ranking is mentioned in Table 14.14.

14.5.2 F-VIKOR Computations

Step 1: The normalized decision matrix for VIKOR was obtained via the same process as TOPSIS and is shown in Tables 14.8–14.11.

Steps 2 and 3: The normalized matrix from the previous step was defuzzified by Equation 14.11 and is shown in Table 14.15 for a 75% load. Next, the defuzzified matrix was again normalized using Equation 14.4. Using 75% load as an example, NO_x was calculated as tracks.

Step 4: The finest and poorest values were identified using Equation 14.12 for smoke at 75% load, as illustration and detailed in Table 14.16.

$$\text{Fuzzy best value} = \{0.00008, 0.00007, 0.00007, 0.00009, 0.00010, 0.00012\}$$
$$= 0.00007$$

$$\text{Fuzzy worst value} = \{0.00008, 0.00007, 0.00007, 0.00009, 0.00010, 0.00012\}$$
$$= 0.00012$$

Step 5: The utility and regret measure were computed using Equations 14.13 and 14.14, and is shown in Table 14.17.

TABLE 14.14
Closeness Coefficient and Ranks

Load	0%		25%		50%		75%		100%	
Blends	CC_i	Rank	CC_i	Rank	CC_i	Rank	CC_i	Rank	CC_i	Rank
Diesel	0.547734	3	0.843199	3	0.588251	3	0.584646	3	0.543993	4
B20	0.616303	1	0.862637	2	0.677414	2	0.716584	1	0.615657	2
B40	0.604646	2	0.87894	1	0.694607	1	0.648128	2	0.670616	1
B60	0.520092	4	0.798837	4	0.563015	4	0.476012	5	0.61473	3
B80	0.4616	5	0.776642	5	0.52031	5	0.529249	4	0.49232	5
B100	0.398407	6	0.140885	6	0.391106	6	0.28795	6	0.393894	6

TABLE 14.15
Defuzzified Fuzzy Decision Matrix for 75% Load

Criteria Blends	NO_X	Smoke	BTE	CO_2	CO	HC	EGT	ID	CD	MRPR
Diesel	1356.167	24.77639	30.56	6.208333	0.088333	36.33333	322.5833	13.505	52.86333	6.193333
B20	1356.167	23.74167	30.56	5.85	0.088333	35.11111	324.4306	13.505	52.86333	5.551111
B40	1350.972	23.74167	30.56	5.991667	0.08	33	329.1806	13.505	48.69667	5.211111
B60	1345.472	28.91528	27.35167	6.208333	0.08	30.88889	326.8056	13.505	49.80778	5.211111
B80	1339.972	28.91528	27.06	6.208333	0.071667	29.66667	331.5556	12.57333	51.68278	4.531111
B100	1337.833	36.15833	27.06	6.35	0.071667	30.88889	338.4167	11.355	50.43278	3.926667

TABLE 14.16
Fuzzy Best (F^*) and Fuzzy Worst (F^-) values

Load	0%		25%		50%		75%		100%	
Criteria	F^*	F^-	F^*	F^-	F^*	F^-	F^*	F^-	F^*	F^-
NO_X	0.00280	0.00288	0.00163	0.00168	0.00095	0.00100	0.00070	0.00070	0.00056	0.00057
Smoke	0.00002	0.00007	0.00005	0.00009	0.00006	0.00010	0.00007	0.00012	0.00011	0.00017
BTE	0.00000	0.00000	0.00006	0.00005	0.00008	0.00007	0.00010	0.00009	0.00011	0.00010
CO_2	0.00001	0.00001	0.00001	0.00001	0.00001	0.00002	0.00002	0.00002	0.00002	0.00003
CO	0.00000	0.00000	0.00000	0.00000	0.00000	0.00000	0.00000	0.00000	0.00000	0.00000
HC	0.00008	0.00009	0.00008	0.00011	0.00009	0.00012	0.00010	0.00012	0.00010	0.00012
EGT	0.00050	0.00056	0.00064	0.00073	0.00084	0.00093	0.00102	0.00116	0.00123	0.00137
ID	0.00005	0.00005	0.00005	0.00005	0.00004	0.00004	0.00004	0.00004	0.00004	0.00004
CD	0.00014	0.00015	0.00014	0.00015	0.00015	0.00016	0.00015	0.00017	0.00016	0.00017
MRPR	0.00001	0.00001	0.00002	0.00001	0.00002	0.00001	0.00002	0.00001	0.00002	0.00001

TABLE 14.17
S_i and R_i Values

Load	0%		25%		50%		75%		100%	
Blends	S_i	R_i	S_i	R_i	S_i	R_i	S_i	R_i	S_i	R_i
Diesel	0.437078	0.161728	0.386072	0.154864	0.541559	0.158161	0.562776	0.161728	0.629676	0.12216
B20	0.370629	0.109946	0.288798	0.161728	0.338925	0.161728	0.461683	0.149229	0.408506	0.161728
B40	0.305669	0.111847	0.302755	0.117652	0.329475	0.111347	0.344404	0.108529	0.374584	0.062181
B60	0.345626	0.060807	0.330726	0.10461	0.489711	0.074396	0.603826	0.137859	0.425256	0.079322
B80	0.425383	0.098672	0.457242	0.132548	0.512595	0.147222	0.530027	0.147222	0.4037	0.079653
B100	0.507642	0.150123	0.658697	0.150123	0.57931	0.150123	0.572724	0.150123	0.584311	0.150123

TABLE 14.18
S^*, S^-, R^* and R^- Values

Load	S^*	S^-	R^*	R^-
0	0.30566	0.50764	0.0608	0.16172
25	0.28879	0.65869	0.1046	0.16172
50	0.32947	0.5793	0.07439	0.16172
75	0.3444	0.60382	0.10852	0.16172

TABLE 14.19
F-VIKOR Ranking

Blends	Q_i	Rank	Q_i	Rank	Q_i	Rank	Q_i	Rank	Q_i	Rank
Diesel	0.40426	3	0.47224	3	0.51891	3	0.72142	3	0.18541	3
B20	0.25286	2	0.05667	1	0.32068	2	0.60856	2	0	1
B40	0.09891	1	0.13303	2	0.21155	1	0	1	0.14482	2
B60	0.48395	4	0.57139	5	0.78343	4	0.77565	4	0.56648	4
B80	0.82531	5	0.5	4	0.90402	5	0.92088	6	0.80126	5
B100	0.9425	6	0.89841	6	0.93355	6	0.83098	5	0.85279	6

Step 6: The maximum and minimum values of utility and regret measures were determined using Equations 14.15 and 14.16 and are tabulated in Table 14.18.

Step 7: Lastly, the VIKOR index was computed using Equation 14.17. A similar methodology was repeated for 25%, 50%, 100% and no-load condition, and the final ranking of the alternatives is listed in Table 14.19.

14.5.3 Intuitionistic Fuzzy MOORA Computations

Step 1: The normalized decision matrix was similar to TOPSIS, as shown in Tables 14.8–14.11.

Step 2: The computation was performed using Equation 14.18 and was ranked with the highest value as best, as depicted in Table 14.20.

Therefore, the multioptimization objective problem shows that B20 is the best blend under 25% and full-load conditions whereas B40 is the best under 25% and half-load condition.

14.5.4 F-COPRAS-G Computations

Step 1: The weighted normalized decision matrix for this method was as the fuzzy TOPSIS technique.

TABLE 14.20
F-MOORA Ranking

Load	0%		25%		50%		75%		100%	
Blends	y_i	**Rank**	y_i	**Rank**	y_i	**Rank**	y_i	**Rank**	y_i	**Rank**
Diesel	–0.33872	3	–0.23962	3	–0.2712	3	–0.26939	3	–0.26848	2
B20	–0.33377	2	–0.2325	1	–0.26156	1	–0.26495	2	–0.26455	1
B40	–0.32926	1	–0.23589	2	–0.26293	2	–0.25909	1	–0.26908	3
B60	–0.34285	4	–0.24625	4	–0.27715	5	–0.2738	4	–0.27462	4
B80	–0.35035	5	–0.25026	5	–0.27404	4	–0.27778	5	–0.28043	5
B100	–0.3598	6	–0.35163	6	–0.29048	6	–0.29268	6	–0.28876	6

TABLE 14.21
P_j and R_j Computations

	0%		25%		50%		75%		100%	
R	P_j	R_j	P_j	R_j	P_j	R_j	P_j	R_j	P_j	R_j
Diesel	0.012337	0.151883	0.061061	0.16778	0.062142	0.171339	0.060053	0.169599	0.058033	0.172273
B20	0.011212	0.154393	0.0611	0.166182	0.061051	0.167624	0.058759	0.166197	0.058324	0.16752
B40	0.011212	0.157347	0.056584	0.16605	0.056553	0.165006	0.057625	0.164763	0.057202	0.16934
B60	0.010204	0.151154	0.054215	0.169526	0.053268	0.170219	0.054151	0.170271	0.055478	0.170286
B80	0.010204	0.147888	0.05109	0.168777	0.050477	0.167413	0.051358	0.170182	0.049585	0.168805
B100	0.009312	0.15817	0.048341	0.197338	0.050285	0.174607	0.04925	0.175576	0.048621	0.17663

Step 2: P_j and R_j are computed using Equations 14.19 and 14.20 and are shown in Table 14.21. The minimum value of R was also calculated using Equation 14.21. A sample calculation for 75% load and B40 blend is shown below:

$$P_j = \frac{1}{2}\left[(0.0161 + 0.06182) + (0.00444 + 0.03289)\right] = 0.05763$$

$$R_j = \frac{1}{2}\begin{bmatrix} (0.0174 + 0.07563) + (0.01283 + 0.0323) \\ +(0.01205 + 0.05938) + (0.01007 + 0.00015) \\ +(0.00808 + 0.00471) + (0.00675 + 0.03881) \\ +(0.00651 + 0.00842) + (0.00533 + 0.03109) \end{bmatrix} = 0.16476$$

Next, the relative weight of the alternative (Q_j) was calculated using Equation 14.22 and is tabularized in Table 14.22. A sample calculation for the B40 blend of 75% load is as follows:

$$Q_j = 0.05763 + \frac{(0.1696 + 0.1662 + 0.16476 + 0.17027 + 0.17018 + 0.17558)}{0.16476\left(\frac{1}{0.1696} + \frac{1}{0.1662} + \frac{1}{0.16476} + \frac{1}{0.17027} + \frac{1}{0.17018} + \frac{1}{0.17558}\right)}$$

$$= 0.23179$$

Step 3: The value of optimal criteria K was computed using Equation 14.23. The utility degree (N_j) value was calculated using Equation 14.24. A sample calculation of B20 blend at 75% load is as follows:

$$N_j = \frac{0.23142}{0.23179} \times 100 = 99.8409\%$$

TABLE 14.22
Q_j Computation

Load Blends	0%	25%	50%	75%	100%
Diesel	0.167331	0.237973	0.229505	0.229247	0.22734
B20	0.163686	0.239713	0.232123	0.231416	0.232434
B40	0.160824	0.23534	0.23034	0.231785	0.229442
B60	0.165945	0.229304	0.221733	0.222677	0.22676
B80	0.169384	0.226957	0.221766	0.219972	0.22237
B100	0.158145	0.198755	0.214517	0.212684	0.213751

A similar process was performed for partial-load, full-load and no-load condition, and the final ranking is detailed in Table 14.23. The utility degree shows that the B40 blend is the best alternative as it is the best blend at 25%, 50% and full-load conditions.

14.5.5 Comparing F-VIKOR Results With Other MCDM Methods

The results obtained for all the techniques are tabularized in Table 14.24. The orders of the blends for fuzzy TOPSIS, fuzzy VIKOR, fuzzy MOORA and COPRAS-G are B100<B80<B60<Diesel<B20<B40, B100<B80<B60<Diesel<B20<B40, B100<B80<B60<Diesel<B40<B20 and B100<B80<B60<Diesel<B40<B20, respectively. The comparisons between the techniques are discussed in detail below in Table 14.24.

14.5.5.1 Fuzzy VIKOR with Fuzzy TOPSIS

The results of F-TOPSIS and F-VIKOR were ranked according to the relative coefficient and the VIKOR index as B100<B80<B60<Diesel<B20<B40. On comparing the two techniques, B40 is the optimum blend. In TOPSIS, the closeness coefficients are not close to the ideal solutions. As an example, for 25% load, B40 is ranked as first with an aggregate of 0.9433(1-0.05667) for VIKOR, which is very close to the ideal value of 1. However, in TOPSIS, B40 has a nearness coefficient rate of 0.87894, which is not close to 1. Similar process was carried out with 0%, 25%, 75% and 100% loads. According to the analysis, fuzzy VIKOR provides valuable assistance in selecting the optimum blend. Therefore, the final ranking is in the order of B100<B80<B60<Diesel<B20<B40.

14.5.5.2 Fuzzy VIKOR with Fuzzy MOORA

Similar to F-VIKOR, F-MOORA rankings are based on the separation measurement. F-MOORA method deals with the variation between the added beneficial criteria values and the nonbeneficial criteria values. The overall ranking depends upon the performance index. The overall performance index obtained in F-MOORA was compared with the F-VIKOR with the blends ranked as B100<B80<B60<Diesel<B20<B40. The calculation timing is very high and complex if the criteria and the alternatives increase.

14.5.5.3 Fuzzy VIKOR with COPRAS-G

In this case, the ranking was done by comparing the straight and proportionate percentage of the best result and the percentage of the ideal worst result. The ranking of the blends was in the order of B100<B80<B60<Diesel<B40<B20. The utility degree was obtained by comparing each alternative with other alternatives. This comparison involved lengthy computations to determine the best and the worst solution, as well as the utility degree. Moreover, F-VIKOR eliminates the disorientations obtained with COPRAS-G.

CONCLUSION

The biodiesel blends produced from fish oil serve as the best replacement, which is abundant in coastal areas. The choice of selecting the right blend of biodiesel by varying fish oil concentration in diesel using the hybrid MCDM method helps achieve

TABLE 14.23
COPRAS-G Ranking

Load	0%		25%		50%		75%		100%	
Blends	N_j (%)	Rank	N_j (%)	Rank	N_j (%)	Rank	N_j (%)	Rank	N_j (%)	Rank
Diesel	97.96963	3	99.27419	2	98.87211	3	98.90485	3	98.5	3
B20	100	1	100	1	100	1	99.84088	2	98.71248	2
B40	98.78755	2	98.1756	3	99.23183	2	100	1	100	1
B60	96.63595	4	95.65778	4	95.5237	5	96.07047	4	97.55881	4
B80	94.94608	5	94.67869	5	95.5381	4	94.90344	5	95.66988	5
B100	93.36463	6	82.91344	6	92.41504	6	91.75904	6	91.96186	6

TABLE 14.24
MCDM Ranking Comparisons

Load	F-TOPSIS					F-VIKOR				
Blends	**0%**	**25%**	**50%**	**75%**	**100%**	**0%**	**25%**	**50%**	**75%**	**100%**
Diesel	3	3	3	3	4	3	3	3	3	3
B20	1	2	2	1	2	2	1	2	2	1
B40	2	1	1	2	1	1	2	1	1	2
B60	4	4	4	5	3	4	5	4	4	4
B80	5	5	5	4	5	5	4	5	6	5
B100	6	6	6	6	6	6	6	6	5	6

Load	F-MOORA					COPRAS-G				
Blends	**0%**	**25%**	**50%**	**75%**	**100%**	**0%**	**25%**	**50%**	**75%**	**100%**
Diesel	3	3	3	3	2	3	2	3	3	3
B20	2	1	1	2	1	1	1	1	2	2
B40	1	2	2	1	3	2	3	2	1	1
B60	4	4	5	4	4	4	4	5	4	4
B80	5	5	4	5	5	5	5	4	5	5
B100	6	6	6	6	6	6	6	6	6	6

the emission standards. Hence, in this research, F-TOPSIS, F-VIKOR, F-MOORA and COPRAS-G were used to identify the best biodiesel blend. As the numerical values are very accurate and distinct, the fuzzy sets are implemented by decision-makers to categorize them in a better manner as it is a multicriteria group choice creation process. As the MCDMs used are rank-based methods, the final results are assembled and sorted. The blends are ranked as B100<B80<B60<Diesel<B20<B40, B100<B80<B60<Diesel<B20<B40, B100<B80<B60<Diesel<B40<B20 and B100<B80<B60<Diesel<B40<B20, respectively, for F-TOPSIS, F-VIKOR, F-MOORA and COPRAS-G based on the comparisons shown in Table 14.24. From the findings, it is observed that the B40 blend stands first for TOPSIS and VIKOR, followed by MOORA and COPRAS-G. The computation timing of the closeness coefficient and the separation measurement in TOPSIS and MOORA become very high and complex if the criteria and the alternatives increase. Similarly, COPRAS-G also involves lengthy computations to determine the best and worst solution as well as the utility degree. Hence, it is observed that F-VIKOR provides valuable assistance in selecting the optimum blend by eliminating the disorientations associated with other techniques. The current research can be extended to Homogeneous Charge Compression Ignition (HCCI) and multicylinder engines to predict the best blend by varying the combination amounts with an interval of 10% to achieve improved precision.

REFERENCES

Archana, M., Sujatha, V. 2012. Application of fuzzy MOORA and GRA in multi-criterion decision making problems. *International Journal of Computer Applications* 53: 46–50.

Aryee, A.N.A., Van de Voort, F.R., Simpson, B.K. 2009. FTIR determination of free fatty acids in fish oils intended for biodiesel production. *Process Biochemistry* 44: 401–405.

Behcet, R. 2011. Performance and emission study of waste anchovy fish biodiesel in a diesel engine. *Fuel Processing Technology* 92:1187–1194.

Boyd, M., Murray, H.A., Schaddelee, K. 2004. Biodiesel in British Columbia-feasibility study report. *WISE Energy Co-OP/Eco-Literacy:* 1–126.

Brauers, W.K.M., Zavadskas, E. K., Peldschus, F., Turskis, Z. 2008. Multi-objective optimization of road design alternatives with an application of the MOORA method. *Proceedings of the 25th International Symposium on Automation and Robotics in Construction*, Vilnius Gediminas Technical University, Lithuania:541–548.

Buyukozkan, G., Guleryuz, S. 2016. Fuzzy multi criteria decision making approach for evaluating sustainable energy technology alternatives. *International Journal of Renewable Energy Sources* 1:1–6

Cavallaro, F. 2010. Fuzzy T OPSIS approach for assessing thermal-energy storage in concentrated solar power (CSP) systems. *Applied Energy* 87:496–503.

Chandrasekhar, V.S., Raja, K. 2016. Material selection for Automobile torsion bar using Fuzzy TOPSIS tool. *International Journal of Advanced Engineering Technology* 7:343–349.

Chang, W., Lu, X., Zhou, S., Xiao, Y. 2016. Quality evaluation on diesel engine with improved TOPSIS based on information entropy. *IEEE Conference on Control and Decision Conference China*:1–4.

Chatterjee, P., Chakraborty, S. 2012. Material Selection using COPRAS and COPRAS-G methods. *International Journal of Materials and Structural Integrity* 6:111–133.

Cristobal, S.J.R. 2011. Multi-criteria decision-making in the selection of a renewable energy project in Spain, The VIKOR method. *Renewable Energy* 6:498–502.

Deng, J. 1982. Control problems of grey systems. *Systems & Control Letters* 5:288–294.

Diakoulaki, D., Zopounidis, D.C., Doumpos, M. 1999. The use of a preference desegregation method in energy analysis and policy making. *Energy* 24:157–166.

Dincer, H. 2015. Profit-based stock selection approach in banking sector using Fuzzy AHP and MOORA method. *Global Research and Economics Research Journal* 4:1–26.

Etghani, M.M., Shojaeefard, M.H., Khalkhali, A., Akbari, M. 2013. A hybrid method of modified NSGAII and TOPSIS to optimize performance and emissions of a diesel engine using biodiesel. *Applied Thermal Engineering* 59:309–315.

Girubha, R.J., Vinodh, S. 2012. Application of fuzzy VIKOR and environmental impact analysis for material selection of an automotive component. *Materials & Design* 37:478–486.

Godiganur, S., Murthy, CH.S., Reddy, R.P. 2010. Performance and emission characteristics of a Kirloskar HA394 diesel engine operated on fish oil methyl esters. *Renewable Energy* 35:355–359.

Jayasinghe, P., Hawboldt, K. 2012. A review of bio-oils from waste biomass: Focus on fish processing waste. *Renewable and Sustainable Energy Reviews* 16:798–821.

Karande, P., Chakraborty, S. 2012. Application of multi-objective optimization on the basis of ratio analysis (MOORA) method for materials selection. *Materials and Design* 37:317–324.

Kato, S., Kunisawa, K., Kojima, T., Murakami, S. 2004. Evaluation of ozone treated fish waste oil as a fuel for transportation. *Journal of Chemical Engineering of Japan* 37:863–870.

Kaya, T., Kahraman, C. 2010. Multi criteria renewable energy planning using an integrated fuzzy VIKOR & AHP methodology, The case of Istanbul. *Energy* 35:2517–2527.

Kundakci, N. 2016. Combined multi-criteria decision making approach based on MACBETH and Multi-MOORA Methods. *The Journal of Operations Research, Statistics, Econometrics and Management Information Systems* 4:17–26.

Li, N., Zhao, H. 2016. Performance evaluation of eco-industrial thermal power plants by using fuzzy GRA-VIKOR and combination weighting techniques. *Journal of Cleaner Production* 135:169–183.

Lin, C.Y., Li, R.J. 2009a. Fuel properties of biodiesel produced from the crude fish oil from the soap-stock of marine fish. *Fuel Processing Technology* 90:130–136.

Lin, C.Y., Li, R.J. 2009b. Engine performance and emission characteristics of marine fish-oil biodiesel produced from the discarded parts of marine fish. *Fuel Processing Technology* 90:883–888.

Liou, J.J.H., Jolanta. T., Zavadskas, E.K., Hshiung, T.G. 2015. New hybrid COPRAS-G MADM Model for improving and selecting suppliers in green supply chain management. *International Journal of Production Research*:114–134.

Madjid, T., Ehsan, M., Nahid, R., Seyed, M.M., Hamidreza, R. 2012. A novel hybrid social media platform selection model using fuzzy ANP and COPRAS-G. *Expert Systems with Applications* 40:5694–5702.

Mayyas, A., Mohammed, A.O., Mohammed, T.H. 2016. Eco-material selection using fuzzy TOPSIS method. *International Journal of Sustainable Engineering* 9:292–304.

Nwokoagbara, E., Olaleye, A.K., Wang, M. 2015. Biodiesel from microalgae: The use of multi-criteria decision analysis for strain selection, *Fuel* 159:2141–2249.

Opricovic, S. 1998. *Multi-criteria optimization of civil engineering systems*, Faculty of Civil Engineering, Belgrade

Perimenis, A., Walimwipi, H., Zinoviev, S., Langer, F.M., Miertus, S. 2011. Development of a decision support tool for the assessment of biofuels. *Energy Policy* 39:1782–1793.

Poh, K.L., Ang, B.W. 1999. Transportation fuels and policy for Singapore: An AHP planning approach. *Computers and Industrial Engineering* 37:507–525.

Preto, F., Zhang, F., Wang, J. 2008. A study on fish oil as an alternative fuel for conventional combustors. *Fuel* 87:2258–2268

Rassafi, A.A., Vaziri, M., Azadani, A.N. 2006. Strategies for utilizing alternative fuels by Iranian passenger cars. *International Journal of Environment Science Technology* 3:59–68.

Rathi, R., Khanduja, D., Sharma, S.K. 2016. A fuzzy-MADM based approach for prioritising Six Sigma projects in the Indian auto sector. *International Journal of Management Science and Engineering* 12:133–140.

Reyes, J.F., Sepulveda, M.A. 2012. PM-10 emissions and power of a diesel engine fueled with crude and refined biodiesel from salmon oil. *Fuel* 85:1714–1719.

Sakthivel, G. 2016. Prediction of CI engine performance, emission and combustion characteristics using fish oil as a biodiesel at different injection timing using fuzzy logic. *Fuel* 183:214–229.

Sakthivel, G., Nagarajan, G. 2011. Experimental studies on the performance and emission characteristics of a diesel engine fuelled with ethyl ester of waste fish oil and its diesel blends. *SAE International Journal of Fuels and Lubricants*:1–6.

Sakthivel, G., Ilangkumaran, M., Nagarajan, G., Raja, A., Ragunadhan, P.M., Prakash J. 2013. A hybrid MCDM approach for evaluating an automobile purchase model. *International Journal of Information and Decision Sciences* 5: 50–85.

Sakthivel, G., Nagarajan, G., Ilangkumaran, M., Gaikwad, A.B. 2014. Comparative analysis of performance, emission and combustion parameters of diesel engine fuelled with ethyl ester of fish oil and its diesel blends. *Fuel* 132:116–124.

Shanmugam, P., Sivakumar, V., Murugesan, A., Umarani, C. 2011. Experimental study on diesel engine using hybrid fuel blends. *International Journal of Green Energy* 8:655–668.

Soufil, M.D., Malekil, M.R.S., Ghobadian, B., Najafi, G., Hashjin, T.T. 2013. The feasibility of using bio lubricants instead of the available lubricants with the help of the multi-criteria decision making software "TOPSIS". *International Journal of Agronomy and Plant Production* 4:2054–2060

Steigers, J.A. 2002. Demonstrating the use of fish oil as fuel in a large stationary diesel engine. *Alaska Energy Authority*: 1–14.

Taylan, O., Kaya, D., Demirbas. A. 2016. An integrated multi attribute decision model for energy efficiency processes in petrochemical industry applying fuzzy set theory. *Energy Conversion and Management* 117:501–512.

Tzeng, G.H., Lin, C.W., Opricovic, S. 2005. Multi-criteria analysis of alternative-fuel buses for public transportation. *Energy Policy* 33:1373–1383.

Uttam Kumar, M., Sarkar, B. 2012. Selection of best intelligent manufacturing system (IMS) under fuzzy MOORA conflicting MCDM environment. *International Journal of Emerging Technology and Advanced Engineering* 2:301–310.

Uygurturk, H. 2015. Evaluating internet branches of banks using fuzzy MOORA method. *International Journal of Management Economics and Business* 11:115–128.

Vatansever, K., Kazancoglu, Y. 2014. Integrated usage of Fuzzy multi criteria decision making techniques for machine selection problems and an application. *International Journal of Business and Social Science* 5:12–24.

Vedaraman, N., Puhan, S., Nagarajan, G., Velappan, K.C. 2011. Preparation of palm oil biodiesel and effect of various additives on NO_x emission reduction in B20: An experimental study. *International Journal of Green Energy* 8:383–397.

Vilela, L., Mata, T.M., Caetano, N.S. 2010. Biodiesel production from fish oil with high acidity. *Third International Symposium on Energy from Biomass and Waste*, Italy:1–14.

Vucijak, B., Kupusovic, T., Midzic, K.,S., Ceric, A. 2013. Applicability of multi criteria decision aid to sustainable hydropower. *Applied Energy* 101:261–267.

Wang, Y.J. 2014. A fuzzy multi-criteria decision-making model by associating technique for order preference by similarity to ideal solution with relative preference relation. *Information Sciences* 268:169–184.

Winebrake, J.J., Creswick, B.P. 2003. The future of hydrogen fuelling systems for transportation: An application of perspective-based scenario analysis using the analytic hierarchy process. *Technological Forecasting and Social Change* 70:359–384.

Yan, G., Ling, Z., Dequn, Z. 2011. Performance Evaluation of coal enterprises energy conservation and reduction of pollutant emissions base on GRD-TOPSIS. *Energy Proceedia* 5:535–539.

Zadeh, L.A. 1965. Fuzzy sets. *Information and Cont*rol 8:338–353.

Zavadskas, E.K., Turskis, Z., Jolanta, T., Marina, V. 2008. Selection of construction project managers by applying COPRAS-G Method. *Reliability and Statistics in Transportation and Communication* 14:462–477.

15 Understanding the Significant Challenges of Software Engineering in Cloud Environments

Santhosh S. and Narayana Swamy Ramaiah
Jain University

CONTENTS

15.1 INTRODUCTION

15.1.1 Cloud Computing Environment

For a long time, when organizations needed to enhance their computational capability, infrastructure capability and storage requirements, they had two options: purchasing the necessary equipment or create tasks that are increasingly proficient in IT. Cloud computing environment is an approach to provide resources to an organization wherein the organization need not worry about the maintenance of the resources. The cloud service provider has the rights and responsibility for maintaining the resources availed by the user [1]. In the cloud computing engineering process, the systematic approach to standardization, commercialization and governance concerns have been discussed [2].

15.1.2 Software Engineering Frameworks

The software engineering-based application and software development has various standards and procedures. As the cloud computing environment has emerged with different standards, languages support and various software development life cycle (SDLC) models, the development process within a cloud environment poses many challenges. Additionally, the cloud provides various services and creates options and perspectives for application development and software development. Hence, programmers can take advantage of the cloud environment, and commercializing the cloud can result in changes in the software industry [3].

The traditional software development life cycle models are working based on the functional parameters and are designed and deployed on the infrastructure and resources available on-premise. The organization has the proper control and better security measures over the infrastructure and resources. As the resources are owned by an organization, the cost of software development and deployment will be the major concern. So the developers and the organizations are interested towards cloud based development as its features are more advantageous than traditional SDLC. On the other hand, its important to understand the challenges of cloud environment while choosing its resources and features as per the requirements of product [4].

15.2 REVIEW OF LITERATURE

The software engineering process based on the cloud platform has benefits such as high availability, agility, flexibility and the overall reduction in the cost of implementation. However, based on the recent concepts of cloud architecture for software development, one can find many issues in the field, including requirement analysis, architectural design, coding and development, testing and quality of service [5] (Table 15.1).

According to Guha [6], cloud computing and web services are the two major paradigm shifts over the conventional form of software development and deployment. These paradigm shifts can make the software development process much more difficult as software developers and programmers are expected to excel in the skills of the semantic web for utilizing the distributed computing environment and the developers must interact with the "cloud provider" during the entire software engineering process [7]. The web services developed for one software can be reused for any other software to reduce the work in terms of kilo lines of code (KLOC) or function points (FPs) on the cloud platform. This can be possible with added skills of semantic web requirements, communication skills and coordination with the cloud service provider, which are considered complex tasks.

According to Raj et al. [8], the high availability and stability of software directly impact customer satisfaction and the company's revenue generation. The software development process is, therefore, an important activity of an organization. Hence, this process is continuously facing major changes in all the phases of SDLC. A growing number of system integrators and independent software vendors transform organizations into cloud providers to deliver applications to the customers and their partners in terms of cloud-hosted services [9]. This environment has a potential to reduce the required time to business services development and to take them to the market. Every service that can be accessible to the user for an additional time directly

TABLE 15.1
Summary of the Literature Review

Sl. No	Title of the Paper	Year	Problems Defined	Problems Resolved	Benefits Obtained	Limitations
1.	Using the Cloud to Facilitate Global Software Development Challenges [10]	2011	a. The software development using global software development (GSD) with the help of cloud computing environment based on the understanding of SOA methods and current GSD method available b. Collaboration challenges	a. To deal with geographic issues, the high availability, runtime automatic adoption, dynamic binding of the services required can help b. The cloud computing environment models consider that the inventory maintained by the centralized location and the data reside in a centralized location where the inventory of services is maintained. Services maintain a registry where all of them are stored. This attribute could be used to store and retrieve configurations	a. The cloud computing environment has the potential of resolving the unsolved hurdles of the GSD projects	Lack of communication between the developers and the cloud service providers in terms of technical requirements
2.	Impact of Semantic Web and Cloud Computing Platform on Software Engineering [6]	2013	a. Need for modification of software engineering process b. Impact of cloud computing on software engineering	The cloud service providers will be able to help in defining the following: (1) Number of programmers required, (2) reusable components, (3) estimation of the cost, (4) scheduling requirements, (5) risk management, (6) quality assurance, (7) change management and (8) configuration management	The amount of reusable components from the previously developed to current web service requirements can be estimated in the number of kilo lines of code (KLOC) or number of function points (FPs) to be newly developed by the software engineer will reduce	a) The complexity of the software product may increase by multiple folds due to poor or no documentation of the implementation details and their integration requirements b) Mastering semantic web

(Continued)

TABLE 15.1 (*Continued*)
Summary of the Literature Review

Sl. No	Title of the Paper	Year	Problems Defined	Problems Resolved	Benefits Obtained	Limitations
3.	Impact of Cloud Services on Software Development Life Cycle [11]	2013	Design for parallelism Design for failure	Achieving data parallelism and task parallelism The analysis for any component failure plays an important role to take a decision against the unexpected behavior of an application	Parallelism significantly improves performance in many areas to achieve more benefits To improve the availability of an application and to ensure the application is working as per the requirements, design as failure plays a vital role	Parallelism may lead to multiple problems which are unexpected or sometimes new problems that introduces more complexity to the project
4.	Cloud-Based Development Using Classic Life Cycle Model [12]	2013	The development of software using the classic waterfall model, the iterative model and other process models [16]	Activities can be taken care of in each phase of classic SDLC. spiral, XP and scrum models	The responsibilities of individual phases of the SDLC process are defined It imbibes the systematic aspect of a waterfall and iterative and scrum prototyping	a. The feasibility study and analysis for the cloud service provider and selection of services is very essential as there are no standards to follow in selecting the cloud services b. The expertise and standards to analyze the cloud-based risks are less in the industry
5.	Impact of Cloud Adoption on Agile Software Development [13]	2013	Agile software development process can be incremental and iterative models	The cloud adoption challenges have to understood by the developing team The benefits and the challenges need to be analyzed and weighed before migration	Agile methodology for the software development using cloud environment provides better control, faster return on investments and improved organization competitiveness	Cloud evaluation for each service plays an important role in determining the cloud providers

(Continued)

TABLE 15.1 (*Continued*)
Summary of the Literature Review

Sl. No	Title of the Paper	Year	Problems Defined	Problems Resolved	Benefits Obtained	Limitations
			The cloud computing environment enables to use agile methodology through ubiquitous nature			The on-premise resources need to be synchronized and the integration needs to be taken care Based on the application, type of security required need to be chosen
6.	Efficient Practices and Frameworks for Cloud-Based Application Development [14]	2013	The Amazon Web Services (AWSs) provide EC2 Instance, Elastic IP Address, Amazon Simple Storage Service (S3) and many more services through its cloud environment Windows Azure provides Windows Azure Compute as an alternate to EC2 in AWS. Specify a role (hosting container) and the number of VM instances	Comparative study of the major cloud service providers in the current industry	The major benefits of the cloud computing environment such as on-demand resources auto-scaling, on-the-go pricing, fault-tolerance and no maintenance can be impactful in proper utilization	There are resources available with more benefits, but people are not utilizing the resources properly yet. So, there is a need for best practices in utilizing cloud and implementing applications on the cloud
7.	Analysis for Security Implementation in SDLC [15]	2014	a) The security-related issues and its risks need to be analyzed in each phase of software development life cycle on the cloud platform	Classification of activities in each phase for security-related issues	Security in SDLC over the cloud Secure web application development life Cycle (SWADLC)	Protecting the information and decision making "Move/not Move" while shifting from one phase to another phase of the life cycle can be a concern to be addressed

impacts revenues and affects the financial statements of an organization. The success of an organization lies in the speed at which an application can be designed, implemented, tested and introduced to the market for customer usage. The impact on the return on investment for any application depends on the time to market, which can be accelerated through a rigorous software engineering process.

According to Chana and Chawla [17], the testing infrastructure for the cloud environment with online access provides the necessary attributes of quality, such as security, scalability, elasticity, performance, availability and reliability. The testing phase of an SDLC needs to be shifted to the cloud environment. The strong reasons of this shift include paradigm shift, computing resource utilization, shorter development cycles, flexibility, reduced cost of development, on-demand resources and access to global markets for customers and providers. In addition, the agile methodlogy supported by an online testing environment provides testing services continuously. Although many organizations are providing unit testing, functional testing and performance testing, few organizations are providing fault-tolerance testing, recovery testing and security testing [18].

Agile software development methods are evolving regularly, and the teams in agile process meet regularly to overcome the communication challenges. Moreover, these team meetings have taken the agile development process to the next level in the field of software development [19].

15.3 CHALLENGES IDENTIFIED

Based on the literature review and analysis, there is a need for a cloud-based SDLC model rather than the traditional SDLC models to get more benefits out of the cloud environment. The major challenges are as follows:

a. Designing a model for a cloud-based SDLC.
b. Identifying the most suitable cloud service provider for the software requirements.
c. Cloud platform evaluation concerning various parameters.
d. Designing a cloud-based testing phase with specially designed test cases for security testing.
e. A collaborative and continuous development environment.

15.4 PROPOSED SOLUTION

The solution is entirely different from the traditional SDLC models available in the present IT industry. Figure 15.1 shows the proposed design of cloud-based SDLC. The model for cloud-based SDLC phases can be designed as follows:

Phase 1: Software requirement specification
Phase 2: Cloud and platform evaluation
Phase 3: Cloud requirements analysis and service level agreement
Phase 4: Design
Phase 5: Implementation
Phase 6: Testing
Phase 7: Maintenance

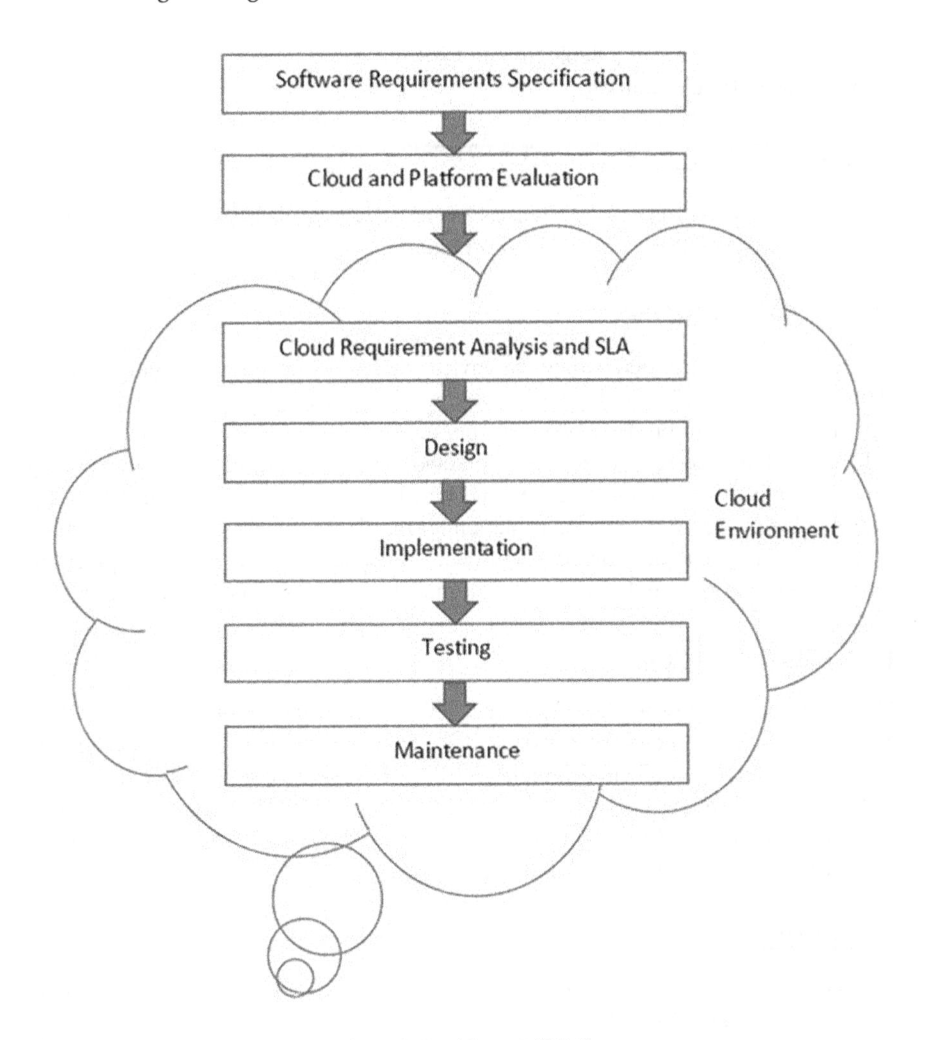

FIGURE 15.1 The proposed design of cloud-based SDLC.

Out of the seven cloud-based SDLC phases defined, Phases 1 and 2 can be carried out outside the cloud environment and phases 3–7 can be carried out within the cloud environment.

15.5 BENEFITS AND LIMITATIONS

15.5.1 Benefits

1. All the benefits of the cloud computing environment such as on-demand services, on-the-go pricing, less maintenance, etc., hold good.
2. Optimized resource utilization after the cloud and platform evaluation as the services are evaluated according to the software requirement specification, which results in cost reduction.

3. Fast development and deployment of software.
4. Reduced time to market.

15.5.2 Limitations

1. Security is an important threat as the software development is done completely on the third-party "cloud provider."

15.6 CONCLUSION

The emergence of the cloud computing environment has brought many changes in the IT industry. The blending of the software engineering process with the cloud computing environment can produce better results in terms of cost reduction and resource utilization. Hence, there is a need to understand the significant challenges in blending the cloud environment and software engineering frameworks. Moreover, the necessary steps need to be taken in devising a solution for the challenges identified. In this chapter, we examined and identified the significant challenges of software development in the cloud environment and proposed a new model to address cloud-based SDLC.

15.7 FUTURE SCOPE

The proposed solution for the cloud-based SDLC focuses on which phases of the process can be implemented outside the cloud environment and which phases can be implemented in the cloud environment [20–22]. However, there are many challenges that are specific to the individual phases of the life cycle. Therefore, there is ample scope to carry forward the work in terms of phase-by-phase evaluation to identify the challenges and address them in the future.

REFERENCES

[1] Grundy, J., Kaefer, G., Keong, J., and Liu, A. 2012, "Guest editors' introduction: Software engineering for the cloud", *IEEE Software*, vol. 29, pp. 26–29.

[2] Shan, T., 7–9 May 2011, Smart cloud engineering, nomenclature, and enablement. In *Proceedings of the 1st International Conference on Cloud Computing and Services Science*, Noordwijkerhout, Netherlands.

[3] Kashfi, H. 2017, "Software engineering challenges in cloud environment: Software development lifecycle perspective", *International Journal of Scientific Research in Computer Science, Engineering and Information Technology*, vol. 2, no. 3, ISSN: 2456-3307.

[4] Goncalves, R., et al. 2015, "A multi-criteria approach for assessing cloud deployment options based on non-functional requirements". *SAC'15 Proceedings of the 30th Annual ACM Symposium on Applied Computing*, pp. 1383–1389.

[5] Silva, E.A.N., and LucreÌ-dio, D. 2012, Software engineering for the cloud: A research roadmap. *SBES-Software Engineering Brazilian Symposium.*

[6] Guha, R. 2013, "Impact of semantic web and cloud computing platform on software engineering". *Software Engineering Frameworks for the Cloud Computing Paradigm, Computer Communications and Networks.* DOI 10.1007/978-1-4471-5031-2_1, Springer-Verlag, London.

[7] Guha R., and Al-Dabass D. 2010, "Impact of Web 2.0 and cloud computing platform on software engineering", *Proceedings of 1st International Symposium on Electronic System Design (ISED).*
[8] Raj, P., Venkatesh, V., and Amirtharajan, R. 2013, "Envisioning the cloud-induced transformations in the software engineering discipline". *Software Engineering Frameworks for the Cloud Computing Paradigm, Computer Communications and Networks*. DOI 10.1007/978-1-4471-5031-2_2, Springer-Verlag London.
[9] Tapangarg, 2010, SAAS, PAAS and IAAS – making cloud computing less cloudy. Available online: http:// cioresearchcenter.com/2010/12/107/
[10] Hashmi, S.I., Clerc, V., Razavian, M., Manteli, C., Tamburri, D.A., Lago, P., Di Nitto, E., and Richardson, I. 2011, Using the cloud to facilitate global software development challenges. *Global Software Engineering Workshop (ICGSEW) 2011 Sixth IEEE International Conference*, pp. 70–77.
[11] Krishna, R., and Jayakrishnan, R. 2013, "Impact of cloud services on software development life cycle". *Software Engineering Frameworks for the Cloud Computing Paradigm, Computer Communications and Networks.* DOI 10.1007/978-1-4471-5031-2_4, Springer-Verlag London.
[12] Balasubramanyam, S.R. 2013, "Cloud-based development using classic life cycle model", *Software Engineering Frameworks for the Cloud Computing Paradigm, Computer Communications and Networks*, DOI 10.1007/978-1-4471-5031-2_5, Springer-Verlag London.
[13] Karunakaran, S. 2013, "Impact of cloud adoption on agile software development", *Software Engineering Frameworks for the Cloud Computing Paradigm, Computer Communications and Networks.* DOI 10.1007/978-1-4471-5031-2_10, Springer-Verlag London.
[14] Anil Kumar M., Pramod, N., and Srinivasa, K.G. 2013, "Efficient practices and frameworks for cloud-based application development", *Software Engineering Frameworks for the Cloud Computing Paradigm, Computer Communications and Networks*, DOI 10.1007/978-1-4471-5031-2_14, Springer-Verlag London.
[15] Raj, G., Singh, D., and Bansal, A. 2014, "Analysis for security implementation in SDLC", *5th International Conference–Confluence the Next Generation Information Technology Summit.*
[16] Bhuvaneswari, T., and Prabaharan, S. 2013, "A survey on software development life cycle models", *International Journal of Computer Science and Mobile Computing*, vol. 2, no. 5, pp. 262–267.
[17] Chana, I., and Chawla, P. 2013, "Testing perspectives for cloud-based applications", *Software Engineering Frameworks for the Cloud Computing Paradigm, Computer Communications and Networks.* DOI 10.1007/978-1-4471-5031-2_7, Springer-Verlag London.
[18] Murthy, N.M.S., and Suma, V. 2014, "A study on cloud computing testing tools". *The 48th Annual Convention of Computer Society of India*, pp. 605–612.
[19] Santhosh S., and Ramaiah, N. 2019, "The impact of software engineering methods for cloud computing models – a survey", *Software Engineering eJournal*, vol 2, no. 17.
[20] Sharp, J., et al. 2014, "Cloud design patterns", *Microsoft Patterns and Practices.*
[21] Federal Bureau of Investigation. 2012, "Recommendations for implementation of cloud computing solutions", *Technical Report of Federal Bureau of Investigation.*
[22] Department of Finance and Deregulation. 2012, "A guide to implementing cloud services", Department of Finance and Deregulation Australian Government.

Index